Cleanroom Microbiology for the Non-Microbiologist

David M. Carlberg, Ph.D.

Interpharm Press, Inc.
Buffalo Grove, IL

10 9 8 7

ISBN: 0-935184-73-2

Interpharm Press, Inc.
1358 Busch Parkway
Buffalo Grove, IL 60089, USA

Phone: + 1 + 708 + 459-8480
Fax: + 1 + 708 + 459-6644

Contents

Dedication

This book is dedicated to my wife, Margaret,
whose patience and understanding made its
writing all the more pleasurable.

Acknowledgment

Many thanks to Ms. Barbara Pogosian,
Golden West College,
whose advice and suggestions during the early stages
of the preparation of this book are very much appreciated.

Preface

Managers, supervisors, plant engineers, maintenance personnel, machine operators, and other workers in biotechnology and bio-engineering areas including pharmaceutical and medical device manufacturers and the food industries are the principal audience for this book. The volume introduces important concepts and principles of microbiology to people who may have little or no training in microbiology but who must deal with problems involving microbial contamination and its control or who regularly interact with microbiologists. While not intended to be a comprehensive textbook on microbiology, the book may also be of some value to trained microbiologists for its overview of cleanroom microbiology.

The objectives of this book are as follows:

- To introduce people who work in clean rooms to the basic vocabulary of microbiology.

- To describe some of the more common types of microorganisms that may be encountered in the clean room, their roles in human activities, and some of the techniques used to study them.

- To explain the techniques used to control microorganisms in the clean room.

- To describe some common types of equipment and techniques used to assess levels of microbiological contamination in the clean room.

- To show the rationales behind many of the regulations and operational procedures associated with biocleanrooms.

- To help establish an understanding relationship between the reader and microbiologist coworkers.

1 The Scope of Microbiology

MICROBIOLOGY is the branch of biology that deals with the smallest living things, the MICROORGANISMS. These organisms include the bacteria, algae, fungi, and protozoa. Viruses, while not living organisms in a strict sense, are also included in the study of microorganisms.

Microorganisms are universal. They nearly always occur wherever other forms of life are present, but they also live in places where environmental conditions are too harsh for other life. Microorganisms are frequently the only living things found at the bottoms of oil wells, for example, or in the hot springs of Yellowstone National Park, in the frozen tundra of Antarctica, and in the extremely dry desert soils of Asia and South America. Bacteria that grow at 110°C (230°F) have been isolated from water near volcanic vents deep in the Pacific Ocean, and there is evidence that others may grow in water as hot as 130°C (266°F).

Natural microbial populations can reach unbelievable levels. Common garden soil is particularly rich in microorganisms. A cubic centimeter (about $1/5$ teaspoon) may contain as many as 1,000,000,000 bacteria. This is a mass that represents 0.3 percent of the total bulk of the soil. Stated another way, an acre of soil down to a depth of 3 feet contains about 3,000 pounds of bacteria.

The human body is also a rich source of microorganisms. As many as a million bacteria may live on each square centimeter of skin, and saliva contains approximately 1,000,000,000 microorganisms per milliliter. Nasal washings from normal, healthy adults have been shown to contain as many as 100,000 microorganisms per milliliter. Fecal matter contains about 100,000,000,000 bacteria per gram, which represents about half its dry weight.

During a normal working day, a person sheds millions of microorganisms into the surrounding environment through breathing, talking, sneezing, coughing, or just moving about. It has been estimated, for instance, that a single sneeze can spray as many as 200,000 microorganisms into the surrounding air. The shedding of dead skin cells from body movement and friction of clothing against the skin also causes the release of enormous numbers of microorganisms. Thus, it is not surprising that soil and people are the two most common primary sources of microbiological contamination in the clean room. Most efforts to control microbial contamination in the clean room are directed toward these sources.

Because microorganisms are known to cause infectious diseases and are common in such large numbers, one might ask why infectious diseases aren't more widespread. Besides the fact that animals, including humans, have developed effective defense mechanisms against infection, the main reason is that of the tens of thousands of species of microorganisms that share the planet Earth with humans and other plants and animals, only a few hundred are known to cause disease; the others are generally harmless and in many instances, beneficial. Many microorganisms, for example, are responsible for maintaining soil fertility, are part of the food chain, or aid in the manufacture of hundreds of products such as wine, vitamins, antibiotics, and fine chemicals. Microorganisms in clean rooms, however, can cause serious problems no matter how beneficial they may be elsewhere. But before we get into the specifics of microbiology, let's first briefly review some fundamental principles of biology.

Cells

A basic principle of biology is that every living organism from the simplest to the most complex is made up of one or more functional units called CELLS. An adult human body consists of billions of cells, each of which has undergone a process known as differentiation. During differentiation, a cell is programmed to carry out a specific function: a blood cell carries oxygen, a muscle cell contracts, a nerve cell conducts impulses, and a gland cell manufactures and secretes hormones. On the other hand most microorganisms, such as bacteria, yeasts, protozoa, and algae, exist as single cells. Each individual microbial cell must carry out all the necessary functions for life: respiration, reproduction, movement, and so forth. The molds, also included in the microorganisms, are multicelled but still considerably simpler than cells in plants and animals. The viruses are in a special classification for they are noncellular infectious particles.

Infection and Disease

Many people think of infection and disease as meaning the same thing, but they do not. Infection means the growth of microorganisms in the body while disease relates to a malfunction of some part of the body. It is possible to have microorganisms growing in the body without creating significant damage, that is, infection without disease. So-called live vaccines such as the Sabin polio vaccine work like that. The viruses grow in the cells of the intestine and stimulate the immune system, but do no harm. However, if activities of the infecting microorganisms begin to adversely affect cells and organs of the host, signs of disease begin to appear. And, of course, one may have disease without infection such as in the case of genetic diseases, which are caused by faulty genetic information.

Binomial Nomenclature

Nomenclature refers to the naming of things. In biology organisms are named according to a system called BINOMIAL NOMENCLATURE, that is, names consisting of two words. The first word represents the GENUS (GENERA, plural), the second word, the name of the SPECIES. Everyone is probably familiar with the scientific name for humans, *Homo sapiens,* for the domestic dog, *Canis familiaris,* or for the horse, *Equus caballus.* A GENUS represents a group of related species. Thus, the genus *Canis* consists of all species closely related to the dog: *Canis lupus,* wolf; *Canis latrans,* coyote; and so on. Scientific names are printed in italics (or underlined) because they are foreign words. The language of binomial nomenclature is Latin, an international language that is understood throughout the world. Thus, a French biologist visiting China would have no trouble explaining what organism is the subject of her research. Every plant, animal, and microorganism is known by the same scientific name around the world even though it may have very different local names.

Occasionally, one may see the term STRAIN used with a particular species of microorganism. The term refers to an organism that differs only slightly from other members of the same species, but not enough to justify placing it into a separate species. For example, there are many strains of the bacterium *Staphylococcus aureus* that are resistant to penicillin, whereas most members of that species are not. If a strain exhibits many differences from other members of the species, but still not enough to justify placing it into another species, it may be referred to as a SUBSPECIES.

Historically, nomenclature has been related to TAXONOMY, the way organisms are related to one another. After genus comes FAMILY, which is a group of related genera. The genera that include bighorn sheep, American bison, domestic cattle, Indian water buffalo, and the musk ox are all distantly related and are thus placed in the same family. The next higher taxonomic levels are ORDER, CLASS, PHYLUM, and finally, KINGDOM. Implied in this classification scheme is the idea that organisms within a given taxonomic grouping evolved from a common ancestor.

The present biological taxonomic system consists of five kingdoms, three of which include the microorganisms. The five kingdoms are ANIMALS; PLANTS; MONERA (containing the bacteria); PROTISTA (containing the algae and the protozoa); and FUNGI. It has recently been proposed that the kingdom monera be split into two kingdoms, EUBACTERIA ("true bacteria") and ARCHAEA ("ancient ones"). Members of the archaea have characteristics of both bacteria and animal cells and appear to represent a very early branching in the evolution of life. Because they appear superficially like bacteria, the archaea were previously called archaebacteria, but because biochemically they do not seem to be any more closely related to bacteria than they are to higher forms, the name was changed. Most of the archaea so far discovered are found in extremely harsh habitats, such as those with high temperatures, high salt concentrations, and extremes in acidity or alkalinity, all conditions that are supposed to have been present in the very early periods of Earth's development. However, many have been found in normal habitats as well.

Taxonomic classifications are based for the most part on physical characteristics, and for nearly all higher plants and animals with their thousands of characteristics, this approach has generally been satisfactory. Such characteristics are so well known that even the average preschooler, for example, can recognize most members of the cat family. When it comes to the taxonomy of microorganisms, however, as you will soon learn, microorganisms are relatively simple creatures with few observable characteristics. This provides microbiologists with a less than satisfactory taxonomic system. Some of the newer techniques of molecular biology will soon make it possible to classify a microorganism by merely reading the genetic information that is encoded in its DNA. Until that time comes, however, microbiologists must be content with the limitations of the present system of microbial classification.

Morphology

MORPHOLOGY refers to the size, shape, and other physical features of an organism. Listing its morphological characteristics is the first step employed in the identification of an unknown microorganism. The unit of length used by microbiologists to measure the size of microorganisms is the micrometer (μm), or micron (μ), a millionth of a meter, which is equivalent to $^1/_{25400}$ of an inch. The remainder of this chapter will be concerned with the morphological characteristics of various microorganisms.

Bacteria

The BACTERIA (BACTERIUM, singular) constitute a large group of thousands of species of single-celled organisms. Because they appear to be the most numerous type of microorganism in soil and on the body, they are probably the most common microbial contaminant in clean rooms. Bacteria occur in three basic shapes (Figure 1.1): rods, or bacilli (bacillus, singular); spheres, or cocci (coccus, singular); and spirals, or spirilla (spirillum, singular). Numerous variations on these three shapes are common (Figure 1.2), such as curved rods, called vibrios, and ovoid spheres, known as coccobacilli.

The arrangements that cocci exhibit in relation to one another also have special names. If the cells occur predominantly in

Figure 1.1. The Three Basic Shapes of Bacteria. All known bacteria can be categorized into one of three basic groups according to their shape.

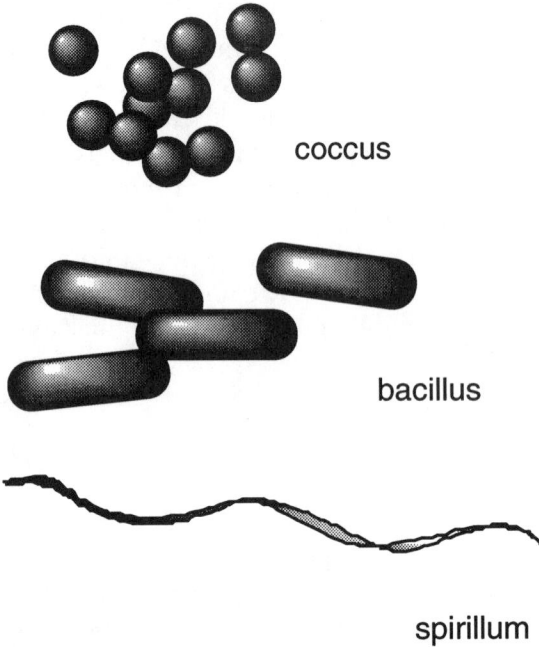

coccus

bacillus

spirillum

Figure 1.2. Variations on Basic Shapes of Bacteria. Many bacteria appear in variations on the three basic shapes, such as coccobacilli (ovoid spheres) and vibrios (curved rods).

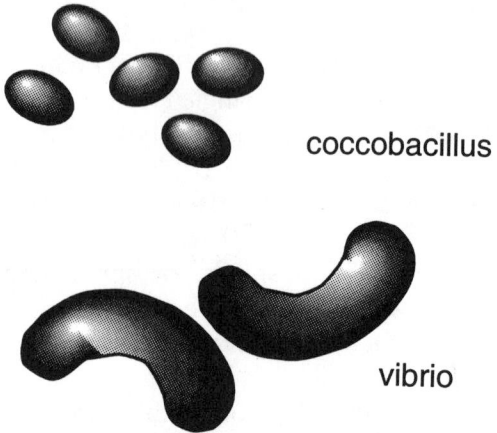

coccobacillus

vibrio

random clusters, they are referred to as staphylococci (from *staphule*, Greek for grapes); if they occur in pairs, they are called diplococci, and they are referred to as streptococci if they are mostly seen in chains (Figure 1.3).

Nearly all the morphological descriptors just introduced have also been used in the formal names of some bacteria: *Bacillus, Staphylococcus, Streptococcus,* and *Diplococcus* are all names of bacterial genera. To determine if a name you read is a specific bacterial genus or merely a morphological descriptor, remember that genus names are always capitalized and in italics, the descriptors are not.

Figure 1.3. Arrangement of Bacterial Cells. Spherical bacteria usually seen in random clusters are called staphylococcus (top). Streptococcus is used to describe cocci that frequently appear in chains (middle). Cocci that form pairs are called diplococcus (bottom).

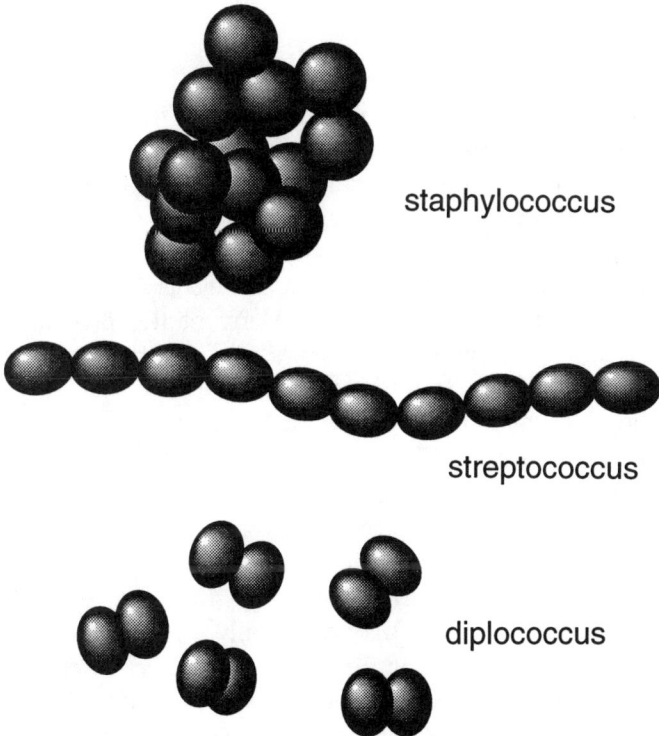

staphylococcus

streptococcus

diplococcus

The size of bacteria depends on the species and the age of the cells. The cells of at least one extraordinary species that was discovered in 1993 reach over 500 μm (micrometers) in length, whereas certain small species may be less than 0.2 μm. Most common bacteria, however, are in the 1 μm to 2 μm size range.

In 1884 the Danish biologist Hans Christian Gram discovered that certain bacteria once stained with a particular dye and treated with a dye-fixing MORDANT would retain the dye even after being subjected to a strong decolorizing solvent such as acetone or alcohol. Other bacteria, on the other hand, would lose the dye, that is, become decolorized on exposure to the solvent. Those species of bacteria that retain the dye are referred to as being Gram positive; those that are more easily destained are Gram negative. The Gram stain, as this procedure is now known, is the most important benchmark in microbiology in the identification of bacteria. Although microbiologists are not entirely sure what is the basis of the Gram stain, the reactions of cells toward the Gram stain appear to be caused primarily by fundamental differences in physical structures of the two types of cells, differences that will be described below.

Figure 1.4 shows some of the features of a typical bacterial cell. The liquid core, or CYTOPLASM of a bacterial cell is surrounded by a thin membrane and an outer, rigid cell wall. Gram negative bacteria have an additional membrane outside the cell wall. The membranes generally control what enters and leaves the cell, and the cell wall both controls the shape of the cell and protects it from external physical forces.

The membranes are made of a double layer of an essentially waterproof, lipid (fatty) material. Many of the proteins that are imbedded in the membranes act as gatekeepers to control the passage of materials in and out of the cell.

The cell walls are composed of many strands of a substance called peptidoglycan. The strands are held together by polypeptide cross-bridges, forming sheets (Figure 1.5). The entire cell wall structure, consisting of many layers of the peptidoglycan sheets, resembles a stack of chain link fences. The cell walls of Gram positive bacteria may consist of 40 or more layers, while Gram negative cell walls may have only a few layers.

The outermost membrane of Gram negative bacteria is made of layers of lipopolysaccharides, phospholipids, and lipoproteins that protect the cell from penetration by harmful substances, such

Figure 1.4. Major Features of a Bacterial Cell.

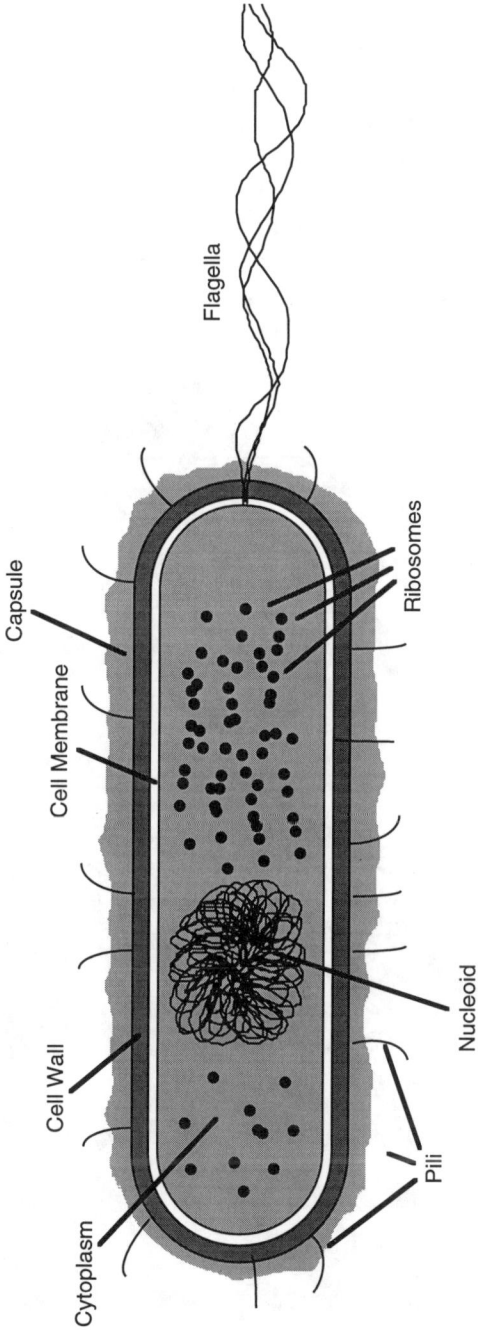

as antibiotics and disinfectants. Gram positive bacteria do not have this outermost protective layer, and are therefore generally more susceptible to these substances. Knowledge of the Gram reaction of a bacterium that may be the cause of an infection is thus helpful in the choice of the type and amount of an antibiotic needed for treatment of the infection. (*Note:* See the glossary at

Figure 1.5. The Bacterial Cell Wall. The cell wall is composed of chains of alternating molecules of acetylmuramic acid (black spheres) and acetylglucosamine (white spheres) that are held together by peptide cross bridges to form sheets (top). Stacks of the sheets make up the cell wall (bottom).

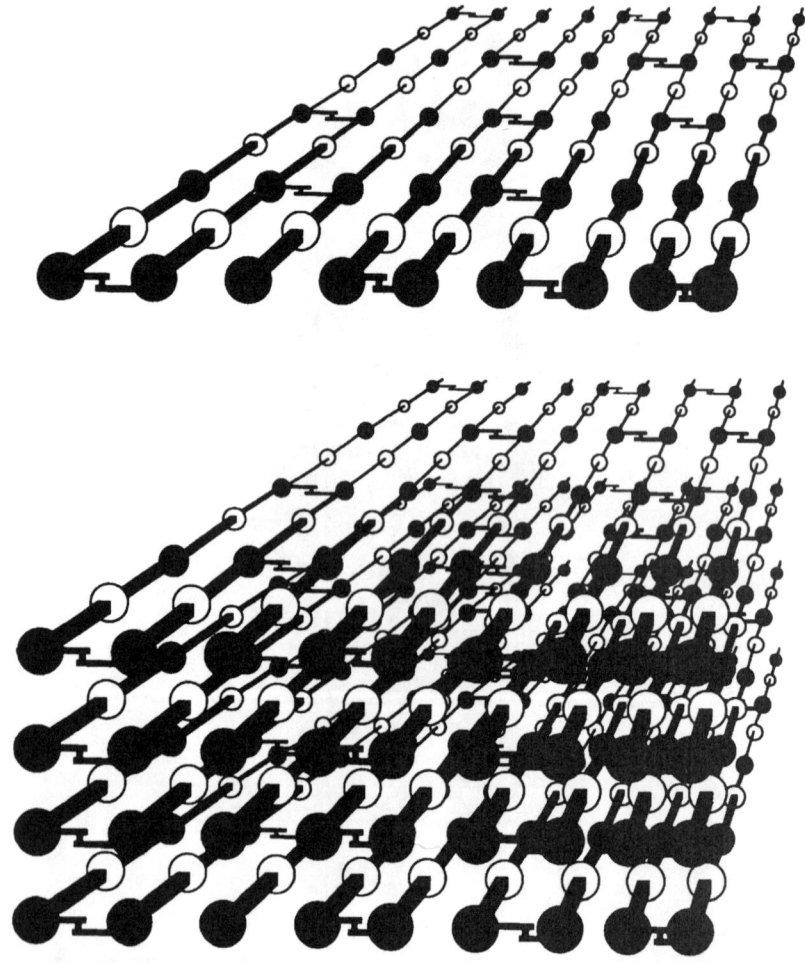

the end of the book for explanations of the chemical and other terms in this section.)

As just mentioned, the outermost layer of Gram negative bacteria contains lipopolysaccharide, a substance that is highly toxic to many animals, including humans, mainly because of its content of a molecule known as lipid A, or endotoxin. Besides being toxic, endotoxin is a pyrogen (fever inducer). When Gram negative bacterial cells die, they frequently disintegrate because of the action of enzymes, causing the release of the endotoxin. If the cells happen to be in the bloodstream of an animal host, the release of endotoxin can cause a severe and even fatal reaction. If the cells or the endotoxins get into pharmaceutical fluids intended for parenteral use, that is, for injection, the effect may be equally dangerous. Therefore, all solutions, instruments, or devices that come in contact with the bloodstream, spinal fluid, or regions of the eye must be essentially free of endotoxin, that is, be pyrogen free.

How do the endotoxins get into parenteral fluids in the first place? Remember, microorganisms are everywhere, including the water and raw products to be used to make up parenteral solutions and other pharmaceutical liquids. Generally, the processes used to purify water, such as distillation, reverse osmosis, and ultrafiltration, rid the water of most impurities, including bacteria and endotoxins. However, unless the water is protected from subsequent contamination, which is extremely difficult if not impossible, bacteria reenter the water mainly through cracks and openings in pipes and storage containers. Although only a few bacteria may gain access to the water supply, certain ones can derive enough nutrients from so-called pure water to multiply to surprisingly large populations. There are usually enough nutrients leaching from plastic tanks and piping, valve lubricants and dust to feed the few bacteria until they reach populations large enough to produce measurable quantities of endotoxin.

As revealed earlier, one of the characteristics of the lipid A endotoxin is its pyrogenicity, or ability to produce fever. This feature has lent itself to a test that is designed to detect the presence of endotoxin in solutions or medical devices. In the Rabbit Pyrogen Assay, test samples are injected into rabbits and the animals are closely observed for the development of fever.

A newer and more sensitive test based on an entirely different principle is the Limulus Amoebocyte Lysate, or LAL, Assay.

This test takes advantage of the fact that the endotoxin causes the gelling of a substance found in the blood cells of the horseshoe crab *Limulus*. The LAL Assay is capable of detecting considerably lower concentrations of endotoxin than the Rabbit Pyrogen Assay, and, of course, it avoids the need for live test animals. The crabs can be captured, partially bled, and returned to the sea without harming them.

External features of bacterial cells

Most species of bacteria are capable of producing large quantities of polysaccharides or polypeptides that accumulate on the outer surface of the cell as a thick, slimy layer called a CAPSULE (Figure 1.6). For some species that cause infections in animals, the capsule plays an important role in protecting the cells from the host's natural defense mechanisms. In other species a thicker fibrous polysaccharide material, called a GLYCOCALYX, can form on the surface of the cell. This substance can act as a cement to anchor cells to hard surfaces such as underwater rocks, the inner surfaces of storage vessels and pipes . . . or one's teeth. Populations of microorganisms attached to surfaces such as these are sometimes referred to as BIOFILMS. Biofilms can cause serious engineering problems such as corrosion and can impede the flow of fluids through pipes.

Some bacteria possess very long, corkscrew-like appendages called FLAGELLA (flagellum, singular) (Figure 1.7), which give a cell the ability to move through liquid. The movement, or motility as it is called, is caused by the propeller action of the rotating flagella generated by molecular-sized motors at the base of the flagella. The motors have bushings, rotors, and stators much like an electric motor, but these motors are powered by a current of protons rather than electrons. Other external appendages, seen only in Gram negative bacteria, are PILI (Figure 1.7). These play a role in the attachment of cells to membrane surfaces of animal hosts, to other bacterial cells, or to the surfaces of products or process equipment.

The bacterial interior

The CYTOPLASM is the fluid that fills the bacterial cell. Found within the cytoplasm are numerous structures and dissolved substances necessary for the growth and reproduction of the cell. The most prominent object in the cytoplasm is the NUCLEOID, the depository of most of the genetic information of the cell. The bacterial nucleoid consists of a single molecule of deoxyribonucleic acid (DNA) (See section on bacterial genetics for more details). In the cells of higher plants, animals, and most of the other microorganisms, the comparable structure (the NUCLEUS) contains several molecules of DNA along with proteins that are part of large and complex structures known as CHROMOSOMES. In

Figure 1.6. The Bacterial Capsule. The capsule of these *Staphylococcus* cells that have just divided appears as the shaggy, dark outer layer. (Electron micrograph courtesy of Tai Wu, California State University, Long Beach.)

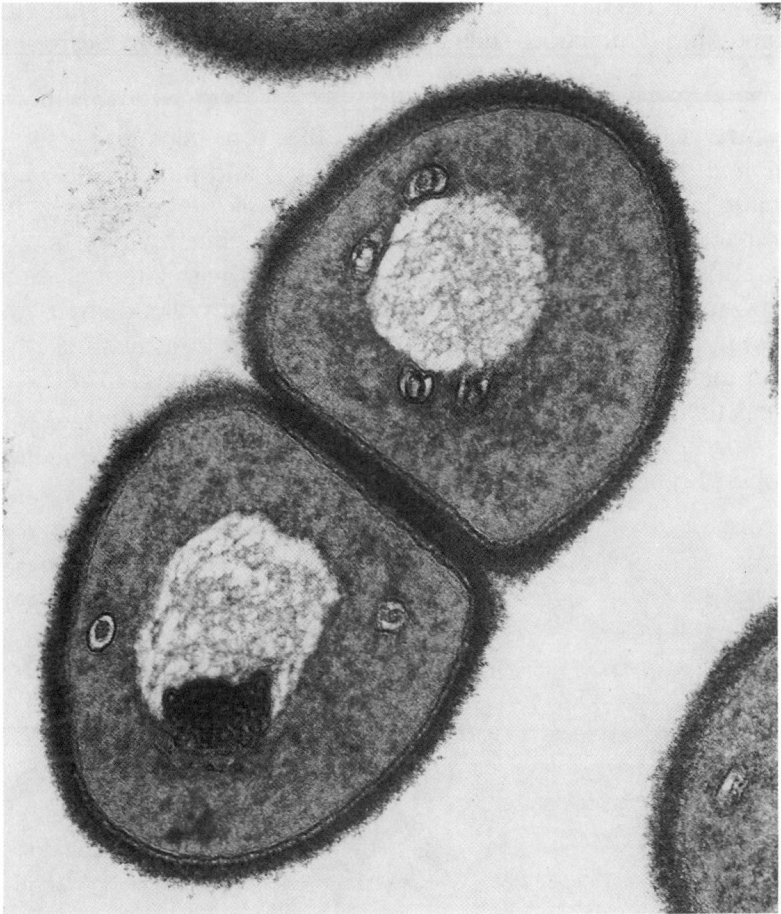

these organisms the nucleus is surrounded by a membrane, the NUCLEAR MEMBRANE. The much simpler nucleoid of bacteria has no surrounding membrane and is therefore in direct contact with the cytoplasm. This is the reason for the difference in names (nucleoid versus nucleus). Consequently, bacteria with their simpler nucleoids, are designated as PROCARYOTES. The cells of nearly all other microorganisms, as well as those of all higher plants and animals, having a true nucleus are called EUCARYOTES. The principal features that distinguish procaryotes from eucaryotes are listed in Table 1.1.

In addition to the nucleoid, many bacteria also possess small fragments of self-replicating DNA called PLASMIDS, which also carry genetic information. While usually not necessary for the growth of a cell, a plasmid frequently can provide the cell with some unique function, such as the ability to digest hydrocarbons,

Figure 1.7. Bacterial Appendages. Electron micrograph of a cholera bacterium showing a flagellum (A) and pili (B). Also appearing are ribosomes (C) that leaked out of the cell during the preparation of the specimen. The shadow effect in this micrograph is achieved by coating the preparation with a thin layer of vaporized platinum. The metal is applied at an oblique angle following which the preparation is observed in a transmission electron microscope. The resulting photograph is printed in reverse; areas that are black are where the platinum is absent and appear as shadows. (Micrograph courtesy of Frederick Eiserling and Robert Romig, University of California, Los Angeles.)

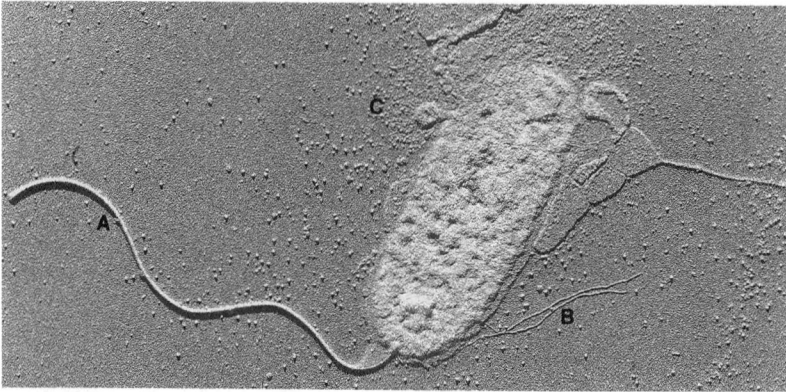

to better penetrate or dismantle host tissue, or to destroy the activity of certain antibiotics.

Plasmids can be transferred from cell to cell through various genetic mechanisms allowing bacteria to share their biochemical capabilities with other bacteria. One mechanism by which a bacterial cell can transfer a plasmid to another cell is CONJUGATION, where the donor cell comes in direct contact with the recipient. In some situations, conjugation may involve the transfer of some donor genetic information, in addition to plasmid DNA. Conjugation is discussed further in the next section on bacterial genetics.

Closer examination of the cytoplasm of bacteria reveals the presence of small particles called RIBOSOMES, which are the protein factories of the cell (Figures 1.4 and 1.7). Copies of nuclear genetic information are made in the form of messenger ribonucleic acid (mRNA). The mRNA amino acids, sources of chemical energy, and other cofactors all congregate at the ribosomes to contribute to the formation of enzymes and other

TABLE 1.1.
COMPARATIVE FEATURES OF
PROCARYOTES AND EUCARYOTES

Features	Procaryotes	Eucaryotes
Cell Wall	Present	Usually absent
Interior membrane system	Absent	Present
Vacuoles	Absent or rare	Usually present
DNA	Usually as a single, circular molecule	Usually in multiple, linear molecules
Histones	Absent	Present
Nuclear Membrane	Absent	Present
Mitotic Apparatus	Absent	Present

proteins. Over 90 percent of the energy consumed by a bacterial cell goes into the synthesis of proteins.

ENDOSPORES are found in a few species of bacteria. The endospore (frequently referred to simply as a spore) is an extremely resistant, dormant body produced primarily by members of the genera *Bacillus* and *Clostridium*. Spores can withstand extreme environmental conditions, such as heat and drying. Some bacterial spores can survive boiling water for several hours and are unaffected by many common disinfectants like alcohol. In the dry state, spores may remain alive for decades, ready to germinate (revert to reproductive vegetative cells) whenever environmental conditions that support growth return. Because bacterial spores are extremely common in soil, any object that comes into contact with soil or dust (which includes just about everything) must be presumed to be contaminated with these resistant forms. Any procedure that is intended to sterilize the object (rid it of all live organisms) must therefore be powerful enough to destroy bacterial endospores.

Bacterial genetics

Whenever plants or animals reproduce, the progeny always look identical or nearly identical to the parents. Camels always have baby camels, and not artichokes, and tuna always have baby tuna and not parrots. It is clear, therefore, that some form of information must be passed on to members of succeeding generations that instruct them how to look like their parents. This transfer of information is called HEREDITY, and GENETICS is the study of heredity.

The genetic information of cells is stored in molecules of DNA, which consist of enormously long strands of subunits called nucleotides strung together like links of a chain. There are four types of nucleotides in DNA, and they are frequently referred to simply as G, A, C, and T. A typical bacterial DNA molecule may be made up of four or five million nucleotides, which if stretched out, would be over a thousand times longer than the cell from which it was removed. A commonly used analogy is if a bacterial cell were the size of a one gallon pickle jar, its DNA would be about 300 meters long. Thus, the DNA in a bacterial cell is like a tightly wound hank of string.

Genetic information is written in a language consisting of only the four nucleotides, but this is more than sufficient to spell out all the instructions that it takes to make a bacterial cell or a human. A typical stretch of genetic information may read

GTGCGAGATGCTGGATGAGCGCCCGATGATCGAGG

The so-called genetic code is the basis for a cell's translation of genetic information into functional protein molecules. The genetic code is universal in that all organisms, from viruses to humans, essentially use the same code. A universal genetic code has made genetic engineering possible. DNA from one organism, when spliced into the chromosome of a cell of a different species, can still be understood and decoded by the recipient organism.

When a cell reproduces, its DNA also reproduces forming identical copies that are distributed to the cell's progeny ensuring that each daughter cell receives a complete and accurate copy of its parent's genetic information. Thus, DNA that we isolate today represents a continuous link back to the beginnings of life. These links to the past are not unaltered, however. DNA experiences MUTATIONS, rare changes in the sequence of nucleotides, which if they are not lethal, are passed on to subsequent generations.

What causes mutations? Background radiation and chemicals in the environment are two possible causes, but the bulk of spontaneous mutations observed in organisms is due to errors cells make when they copy or repair their own DNA.

Bacteria can transfer genetic information from cell to cell by three mechanisms. They are called transformation, conjugation, and transduction. TRANSFORMATION is the transfer of genetic information from cell to cell by naked DNA. Transformation can be accomplished in the laboratory by extracting the DNA from a culture of bacteria and mixing it with a second suspension of live bacteria, usually of the same species. If the recipient cells are competent, that is, physiologically capable of taking up the DNA, they may incorporate the donor DNA into their own genome and soon express one or more characteristics of the bacteria from which the DNA was extracted, such as antibiotic resistance. The recipient cells are said to have become transformed.

The process of incorporating donor DNA into recipient DNA is known as RECOMBINATION, one of the most important processes in all of biology and one of the driving forces in evolution. Recombination occurs when a strand of recipient DNA breaks and allows a strand of donor DNA to become spliced into it.

CONJUGATION requires direct cell-to-cell contact for DNA transfer. Donor (male) cells attach to recipient (female) cells at which time a mechanism is triggered that causes the transfer of donor DNA to the recipient cell. Only a few strains of bacteria are capable of conjugation. The presence of a fertility (F) plasmid is usually necessary for a bacterial cell to act as a donor in conjugation.

TRANSDUCTION involves viruses, called BACTERIOPHAGES, that infect bacteria. As the bacteriophages develop in the host cells (see section on viruses), small bits of host bacterial DNA may inadvertently become packaged into one or more of the virus particles. When the infected bacterial cell bursts, the released virus particles are free to infect other bacteria. If a virus carrying some bacterial DNA infects another bacterium, it will bring bacterial DNA into the cell rather than viral DNA. The recipient bacterium will show no ill effects from this encounter with the bacteriophage for the virus carries little or no viral DNA and is thus incapable of establishing a lethal infection. Again, through the process of recombination, the recipient bacterium may then express the new bacterial genetic information that the bacteriophage delivered to it and, thus, becomes transduced.

It appears that these various manners of genetic transfer probably occur in natural populations of bacteria. Since their discoveries from the late 1920s to the early 1950s, they have been enormously valuable research tools to help biologists learn more about the basic mechanics of genetics. In fact, more has been learned about heredity through the study of microorganisms than through the study of all other plants and animals combined.

Algae

The algae are somewhat distinctive among microorganisms in that they are capable of converting light into chemical energy. That is, they are like plants—able to carry out photosynthesis. Photosynthesis is also found in a few genera of bacteria. To many biologists, the classification of the algae seems in chaos, for some algae closely resemble protozoa and others almost could be classified as plants. Morphologically, algae span nature's extremes of size and complexity. Some algae are nearly as small and simple as bacteria, and, in fact, recently the group known as the blue-green algae was reclassified as bacteria. Other organisms classified as algae, the kelps or seaweeds, may be hundreds of feet long. Since algae are aquatic and photosynthetic, they are always found in environments where ample water and light are available and, thus, will probably not constitute a contamination problem in the clean room except in certain types of circulating water baths, for example.

Fungi

The FUNGI occur in two forms: MOLDS and YEASTS. The appearance of molds is all too familiar to anyone who has kept food unrefrigerated for too long a time. The green or gray fuzzy mass that appears is the MYCELIUM, which is a tangle of multicellular filaments called HYPHAE that is characteristic of the molds (Figures 1.8A and 1.8B). The hyphae may branch, and at their tips occur complex structures that produce SPORES. These spores are specialized cells that serve several functions for the molds: reproduction, dissemination, and protection.

Unlike the case with spore-forming bacteria, where only one spore is formed per cell, a single mold hypha may produce thousands of spores, each of which can be launched into the air with the slightest air movement. If a mold spore happens to land on a site where there are nutrients and moisture, it will germinate and eventually develop into a new mycelium.

While mold spores are somewhat more resistant to harsh conditions than the cells that make up the hyphae, mold spores generally are nowhere near as resistant to adverse environmental conditions as are bacterial endospores. Thus, conditions that destroy bacterial spores easily eliminate mold spores.

Molds are the second most common microbial contaminants in the clean room, and under certain circumstances they can become a serious threat. Their nutritional requirements are generally simpler than those of bacteria, and they can grow in the absence of significant amounts of moisture. Molds are capable, for example, of growing on electronic circuit boards or in air-conditioning filters, deriving their nutrients and water from traces of residual organic contamination and condensed moisture. There is sufficient organic residue, for instance, in a single fingerprint or a small droplet of saliva to support a significant amount of fungal or other microbial growth.

People are most familiar with molds because of their association with food spoilage and the deterioration of materials and equipment through mildew and dry rot. In addition, a few species of molds are responsible for diseases in animals and plants. The destruction of food crops by molds causes multimillion dollar losses each year. On the positive side many of our most effective antibiotics, such as penicillin, griseofulvin, and gentamicin, are produced by molds. Also, many species of molds are necessary in the manufacture of important products, such as corticosteroids and citric acid, and dozens of foods such as soy sauce, miso, and blue cheese.

The second form of fungus are YEASTS, which are single-celled organisms frequently spherical or ovoid, 5 μm to 8 μm in

Figure 1.8A. Molds. The tangled hyphae of the mold *Penicillium*. Bar represents 10 μm.

diameter. Yeasts are capable of forming four or more moderately resistant intracellular spores as one means of reproduction, but more commonly they reproduce by forming BUDS (Figure 1.9). Certain species of yeasts play enormously important roles in the manufacture of many foods and industrial products, such as alcoholic beverages, bread, and vitamins. Because they require high levels of nutrients and water to grow, yeasts are rare contaminants in the clean room. They may, however, grow in certain

Figure 1.8B. A scanning electron micrograph of the conidia and spores of the mold *Aspergillus*. Bar represents 1 μm. (Photo courtesy of Visuals Unlimited/©David M. Phillips.)

types of pharmaceutical products that contain high levels of water and carbohydrates.

While yeasts and molds are generally considered two distinct forms of the fungi, one single celled and one multicelled, a few species are capable of exhibiting both forms at some time during their life cycle or under certain nutritional conditions. These are known as DIMORPHIC fungi.

Figure 1.9. Yeasts. Scanning electron micrograph of *Candida albicans* undergoing budding. Bar represents 1 μm. (Photo courtesy of Visuals Unlimited/©David M. Phillips.)

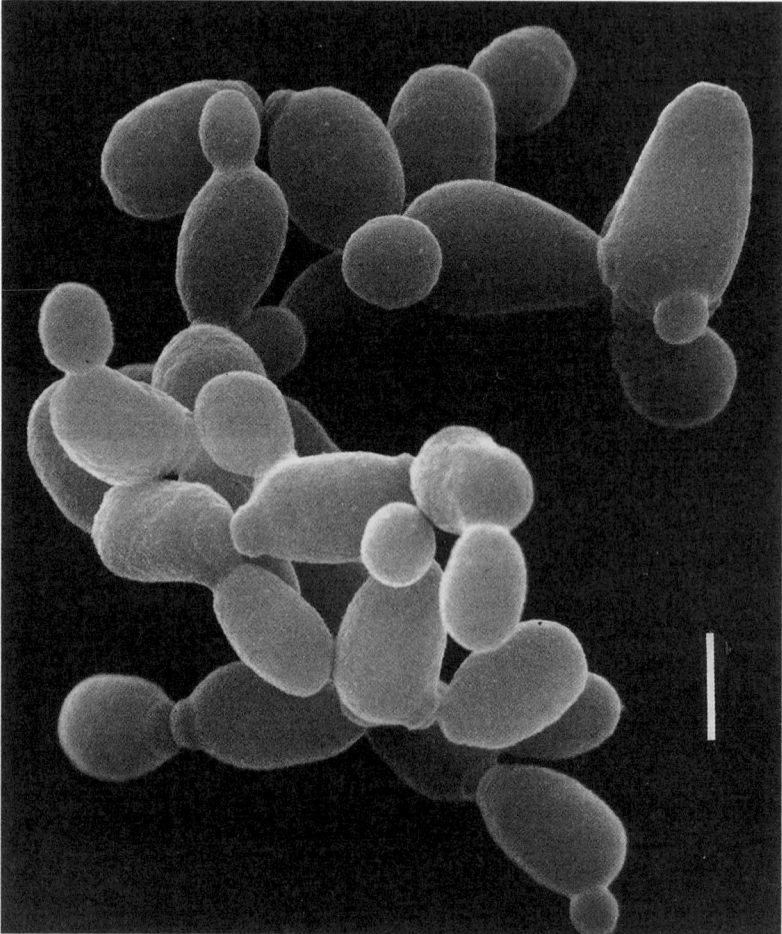

Protozoa

Over 40,000 species of PROTOZOA are known. Most are found free living in natural water environments such as lakes, rivers, and wetlands. The importance of their role in nature as a link in the food chain cannot be overstated. These single-celled, rather complex organisms (Figure 1.10) feed on bacteria and other small microorganisms and, in turn, become food for larger organisms. While only about a dozen genera of protozoa cause diseases such as malaria and African sleeping sickness in humans, in total number of cases worldwide they probably surpass all other infectious organisms combined. Because protozoa are nearly always associated with natural water environments, they are of little concern in cleanroom contamination control.

Figure 1.10. Protozoa. The simplest of the protozoa, *Amoeba,* is still considerably larger and more complex than bacteria. Amoebas are shown in this photograph ingesting small microorganisms and other particles by surrounding them with tentacle-like pseudopods before they are eventually engulfed by cytoplasm. Bar represents 100 μm. (Micrograph courtesy of Visuals Unlimited/©John D. Cunningham.)

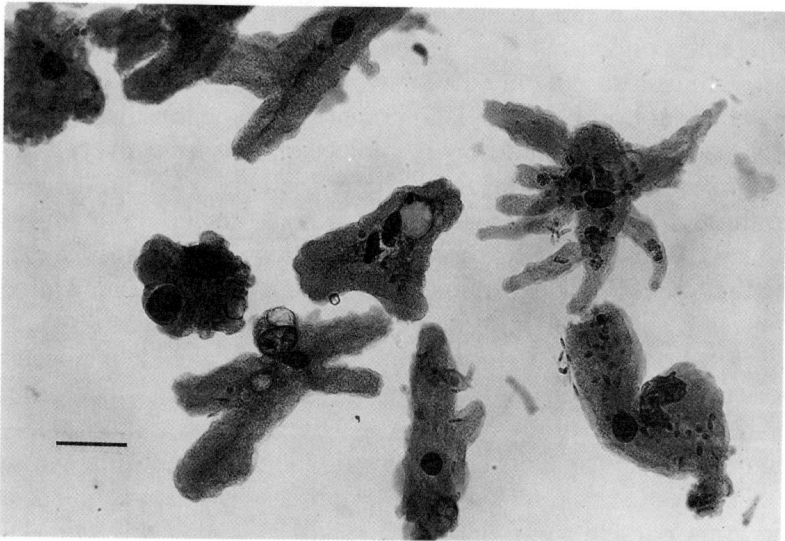

Viruses

As a group the viruses are unique among the microorganisms in that they lack any structures that we normally associate with cells. That is, they are noncellular infectious particles. Their structure is extremely simple, frequently consisting of nothing more than a single molecule of either DNA or RNA (but never both) wrapped with a protein coat, called a CAPSID. Certain more complex viruses may be surrounded by one or more layers of host-derived membranes called ENVELOPES, or may possess specialized structures such as spikes, tails, or fibers. Some viruses contain one or more enzymes needed to infect their host cells. Figure 1.11 shows some typical viral structures. The complete virus particle is known as a VIRION. Table 1.2 lists some common viruses and their dimensions. Clearly, viruses are the smallest of the microorganisms.

Viruses are obligate intracellular parasites, which means they must find an appropriate host cell in which to multiply. To grow viruses in the laboratory, suitable host cell cultures must be maintained. The development of cell cultures for the cultivation of animal and plant viruses is one of the most important advances in microbiology in this century. The cultivation of viruses will be covered in chapter 2. It should be noted that obligate intracellular parasitism is not a unique feature of viruses alone, for many other microorganisms, such as certain species of bacteria and protozoa, must also multiply inside living host cells.

There are probably no living organisms on earth that are free of viruses. Viruses are known to infect organisms at every taxonomic level, from bacteria through the protozoa, algae, fungi, and up through all the higher plants and animals, including of course, humans. Viruses usually display rather narrow host ranges, meaning that, with a few exceptions, a particular virus may only infect a specific species of host, or perhaps a narrow range of related species. For example, viruses that infect fish would not be expected to infect humans, and vice versa.

Viruses replicate in their host cell by synthesis and assembly. Figure 1.12 depicts the steps involved when a typical virus infects a cell. The first step in the infection process is the attachment of the virus particle to an appropriate host cell. It is this step that controls host specificity, for the virus can only adsorb to receptor sites on the host cell surface that match its receptors. Penetration of the host cell by the virus can occur through a variety of

Figure 1.11. Examples of Virus Shapes. Top micrograph shows the polygonal φX174 and tadpole-shaped T$_4$ bacteriophages and the rod-shaped plant pathogen tobacco mosaic virus. Bottom micrograph shows virions of polyoma virus, which infects small rodents. The small spheres that appear to make up the outer structure of the virus are protein subunits called capsomeres; there are 72 to each virion. Bars represent 100 nanometers (nm) (1 nm = 0.001 μm.) (Micrographs courtesy of Frederick Eiserling, University of California, Los Angeles.)

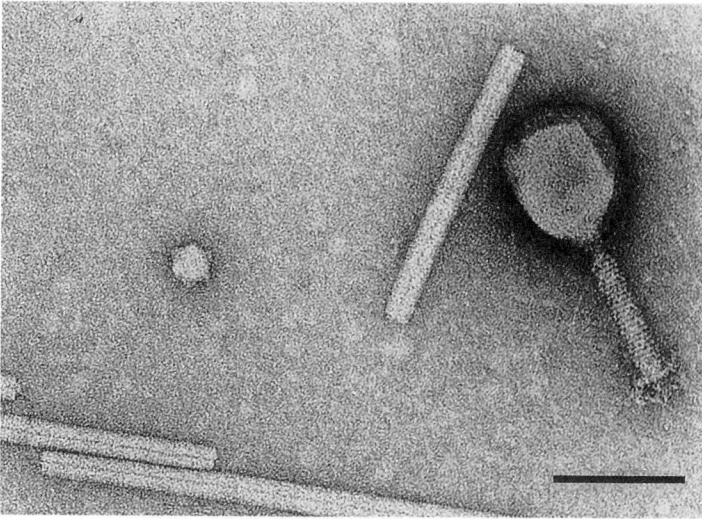

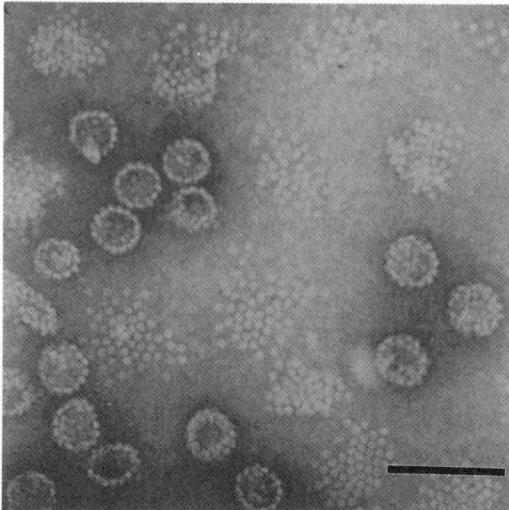

mechanisms depending on the type of virus, but the result is that either the entire virion or just the nucleic acid gains entrance.

If the entire virion enters the host cell, it must be uncoated before the next step in the infectious process can proceed. Once uncoating is completed, the free nucleic acid immediately begins directing the biosynthetic mechanisms of the cell to stop making most host cell-associated molecules and begin making virus nucleic acid, enzymes and virion structure. Soon a stockpile of virus components appears in the host cell, and eventually whole virions begin to accumulate. With certain viruses only a handful of virions may be assembled; in other instances, the host cell may be filled with thousands of virions. The time interval necessary to see this process to its completion may vary from less than twenty minutes in the case of some bacteriophages to many hours for most animal viruses.

The release of the assembled virions from the host cell is the final step in the infectious process. The host cell may simply burst as the result of the destructive activity of an enzyme produced by the virus and release the entire brood of virions at one time. In other instances viruses may be extruded one by one over a period of many hours or days.

In some virus infections one may encounter a LATENT INFECTION in which few, if any, virions are actually manufactured

TABLE 1.2.
SOME VIRUSES AND THEIR DIMENSIONS

Virus	Dimensions (nanometers)
Tobacco Mosaic	17.5 × 300
Herpes simplex	100 (180–200)**
Papilloma (warts)	55
Polio	28
Influenza	9 (90–100)**

*1 nanometer = 0.001 micrometers

**Figures in parentheses include envelope.

From Dulbecco, R., and H. S. Ginsberg. 1988. *Virology*, 2nd ed. Philadelphia: J. B. Lipincott Co. Used with permission.

Figure 1.12. Life Cycle of a Virus. A bacteriophage infects a bacterial host (A) by attaching to its outer surface and injecting its DNA into the bacterial cell (B). The phage DNA undergoes replication, and at the same time enzymes made by the phage DNA cause destruction of the host DNA (C). Within a few minutes the phage DNA begins to direct the synthesis of phage components, such as head and tails, as well as more DNA (D). Assembled mature phage particles begin to accumulate in the host cell (E), and soon enzymes made by the phage lead to the total destruction of the host bacterium and the release of mature phage particles (F). The entire process with bacterial viruses takes less than 20 minutes while animal viruses may require several hours or days to complete an infection cycle.

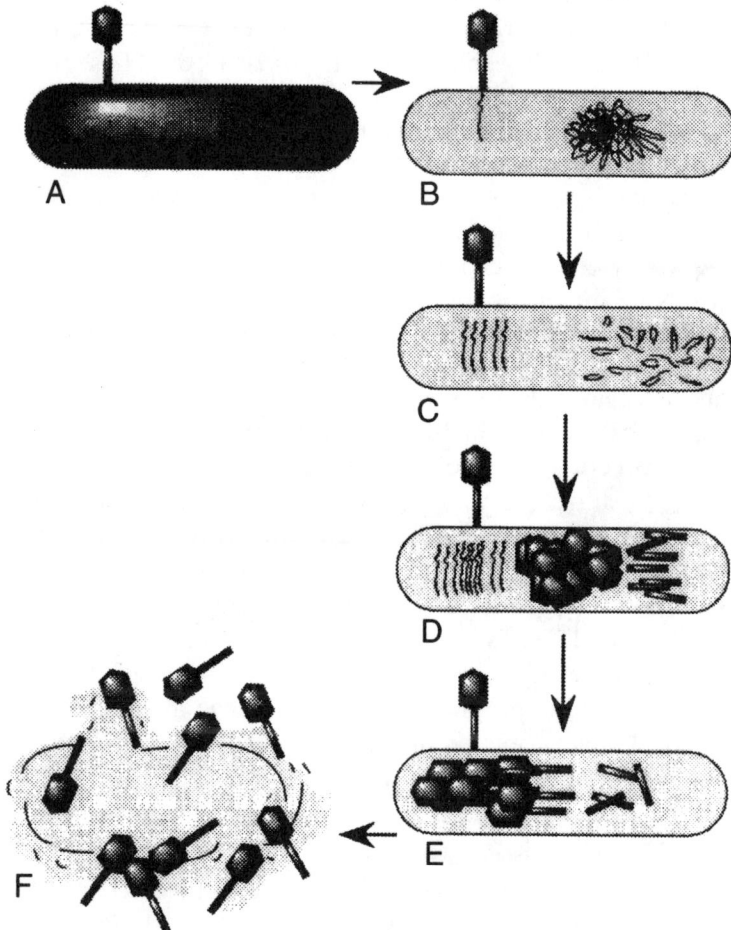

initially. Instead, the nucleic acid of the virus becomes integrated into the DNA of the host cell. Herpes virus is an example of such a case. This virus can remain dormant in host cells for years emerging without warning to initiate active disease with the resulting painful blisters and other typical symptoms. The mechanism of inducing such a dormant state to full blown disease is essentially unknown in most cases. In herpes it appears to be stimulated by various forms of stress, such as fever, bright sunlight, or psychological trauma.

A special group of viruses is known to cause various forms of cancer in animals, including humans. These are the ONCO-VIRUSES, and they can be either DNA or RNA viruses. A common feature of the oncoviruses is their ability to establish a latent infection similar to that described for herpes virus. If the virus happens to contain RNA, an intermediate DNA molecule is synthesized before its integration into the host DNA. These viruses are known as RETROVIRUSES, meaning they make DNA from an RNA template reversing the normal route in which cells make RNA from DNA templates.

Microscopy

Under normal conditions the unaided human eye is probably not capable of seeing objects smaller than about 100 μm. Thus, individual microbial cells are invisible to us without the aid of some type of optical device. Such a device, invented some time in the seventeenth century, is the MICROSCOPE. The familiar magnifying glass is referred to as a SIMPLE MICROSCOPE, consisting of a single lens, whereas a COMPOUND MICROSCOPE consists of two or more lenses. The compound microscope is the type most often seen in microbiology laboratories (Figure 1.13).

MAGNIFICATION is defined as the degree to which an optical system can increase the size of an object's image. While microscopes could be designed to enlarge images millions of times, the practical limit is only around 1000–2000 times. This is because of another limiting characteristic of optical devices, RESOLU-TION. Resolution is defined as the ability of an optical device to separate two very closely spaced objects, that is, to render fine detail.

While resolution of a microscope heavily depends on the quality of its optical components, the wavelength of the light used to illuminate the object under examination also plays a major role. This is due to a fundamental law of physics that one cannot see an object that is about the same size or smaller than the wavelength of the light illuminating it. Common laboratory microscopes use visible light as a source of illumination, the

Figure 1.13. A Light Microscope.

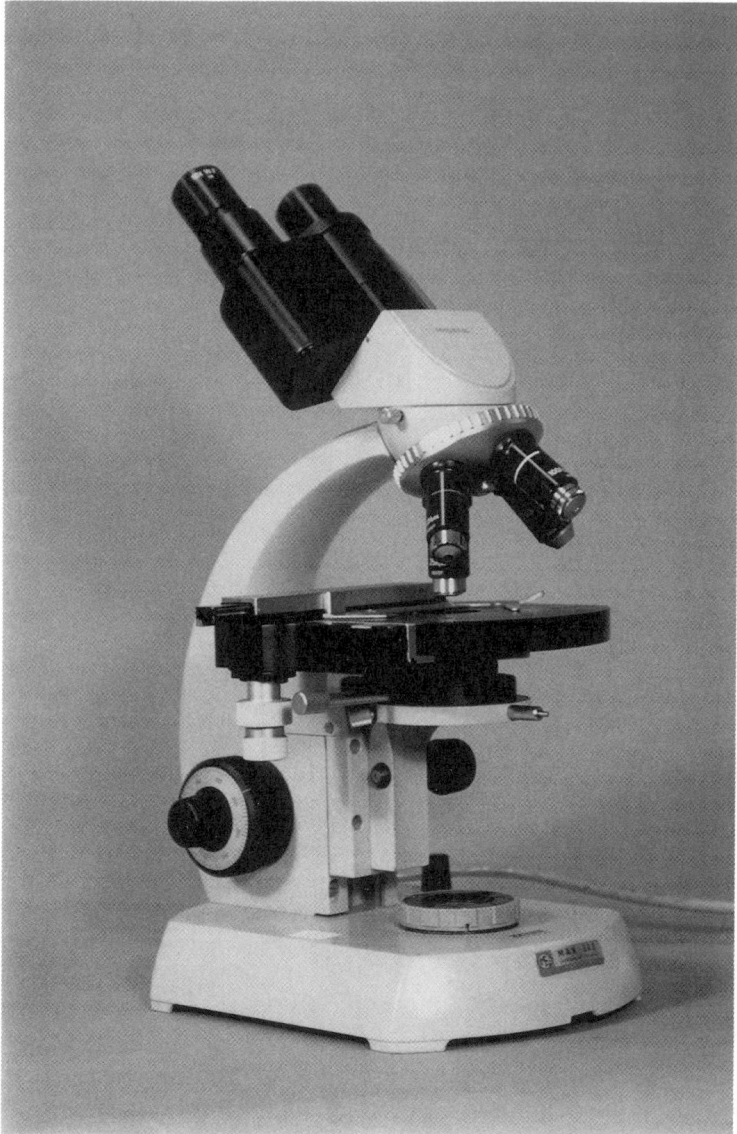

wavelengths of which are only slightly smaller than bacterial cells.

Light Microscopy

Because common microscopes used in a microbiology laboratory use visible light, they are referred to as LIGHT MICROSCOPES (Figure 1.13). The wavelengths of visible light, light that the human eye is sensitive to, range from about 400 nm (nanometers) (deep violet) to 700 nm (deep red). Most bacteria measure around 1 μm or 1000 nm, and their fine structural features, cell walls, flagella, ribosomes, and so on, are considerably smaller. Thus, because the dimensions of bacterial cells are so very close to the wavelengths of visible light, the cells are on the edge of the resolving power of the light microscope, and their fine structure is for the most part beyond its resolving power—it is not visible. Because of their even smaller sizes, most of the viruses are also completely invisible in the light microscope.

Making matters worse, because microbial cells are composed of around 70 percent water, they appear transparent and, thus, only barely visible in the light microscope. Considerable improvement in the visibility (but not resolution) of cells in the light microscope can be achieved by STAINING, which is the application of dyes to impart color to the cells. Staining has the added advantage of giving one the ability to distinguish taxonomic groups of cells, such as with the Gram stain in the case of bacteria. Staining can also improve the visibility of structural features such as nucleoids, endospores, and flagella.

To stain bacteria, a SMEAR is first made by applying a film of cells upon a glass microscope slide. The cells are then FIXED by applying gentle heat to the slide, causing them to stick. The final step is staining, the application of one or more dyes to the smear. If a single dye solution is applied, the process is referred to as a SIMPLE STAIN. DIFFERENTIAL STAINING consists of two or more dye solutions applied one after another, with various washes and other chemical treatments in between the staining steps. Another technique that uses fluorescent dyes is described in chapter 2. Table 1.3 lists the most common differential stains used in microbiology.

Many design improvements in light microscopes have helped to increase the visibility of microorganisms. One is called PHASE

CONTRAST, which takes advantage of the slight difference in index of refraction between the cells and the surrounding medium. Index of refraction is a measure of how much a substance causes a beam of light to bend and is, for example, the basis for the sparkle in diamonds. Another advance is DARK FIELD microscopy, which is based on the ability of microbial cells to scatter light (see chapter 3). Light is focused on the bacterial cells at an angle so the cells appear as bright objects against a dark background.

Electron Microscopy

Resolution can be improved enormously by illuminating objects with a beam of electrons rather than with ordinary light. That is what is done in the ELECTRON MICROSCOPE (Figure 1.14). However, a beam of electrons, which possesses a wavelength considerably shorter than ordinary light, has certain characteristics that make electron microscopy much more difficult to carry out than light microscopy.

Because of collisions with air molecules, a beam of electrons cannot pass through air without severely reducing the intensity of the beam. Consequently, the path of the electrons must be in a high vacuum requiring the specimen under observation to be

TABLE 1.3.
SOME COMMON DIFFERENTIAL STAINS

Stain Technique	Applications
Gram	Divides most bacteria into two groups: Gram positive and Gram negative
Acid Fast	Helps identify members of the genus *Mycobacteria*
Spore	Causes bacterial endospore to stand out in contrasting color
Nuclear	Allows visualization of nuclear region
Flagella	Coats flagella with thick layer of dye so they are visible in the light microscope

dried, a procedure that can lead to shrinkage and distortion of its fine structure. If electrons cannot penetrate air to any extent, then it follows that they cannot penetrate whole cells either. For TRANSMISSION ELECTRON MICROSCOPY, specimens must be prepared in thin sections or slices 1 μm or less to allow passage

Figure 1.14. An Electron Microscope. Notice the size of the instrument compared to the size of the office chair.

of the electron beam. Because there is no way one can hold an individual cell while cutting it into several slices, the cell must be embedded in a plastic material, which is then subjected to sectioning by a glass or diamond knife on an apparatus called a MICROTOME.

To improve contrast various staining procedures involving heavy metals are frequently applied to the sections (Figure 1.15). If one is interested in producing a three dimensional effect, the technique of shadowing (Figure 1.7) is useful.

Although somewhat lacking the high resolution of transmission electron microscopy, SCANNING ELECTRON MICROSCOPY also gives one a three-dimensional impression of cells (Figure 1.16). In this instrument as the electron beam scans the object, it causes a shower of secondary radiation to be released from its

Figure 1.15. Staining in Electron Microscopy. Transmission electron micrograph of a thin section of *Bacillus* cells stained with uranium. The outermost dark layer is the capsule; the light middle layer is the cell wall; and the innermost dark layer is the cell membrane. Note also the central nucleoid. Bar represents 0.5 μm. (Electron micrograph courtesy of Tai Wu, California State University, Long Beach.)

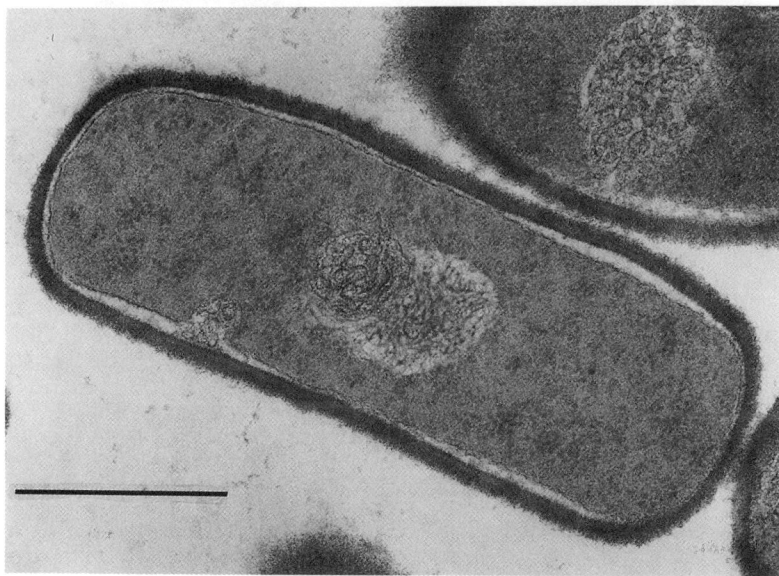

Figure 1.16. Scanning Electron Microscopy. *Staphylococcus* (top) and *Bacillus* (bottom) cells trapped on a membrane filter and observed by scanning electron microscopy. Bars represent 1 μm. (Electron micrograph courtesy of Tai Wu, California State University, Long Beach.)

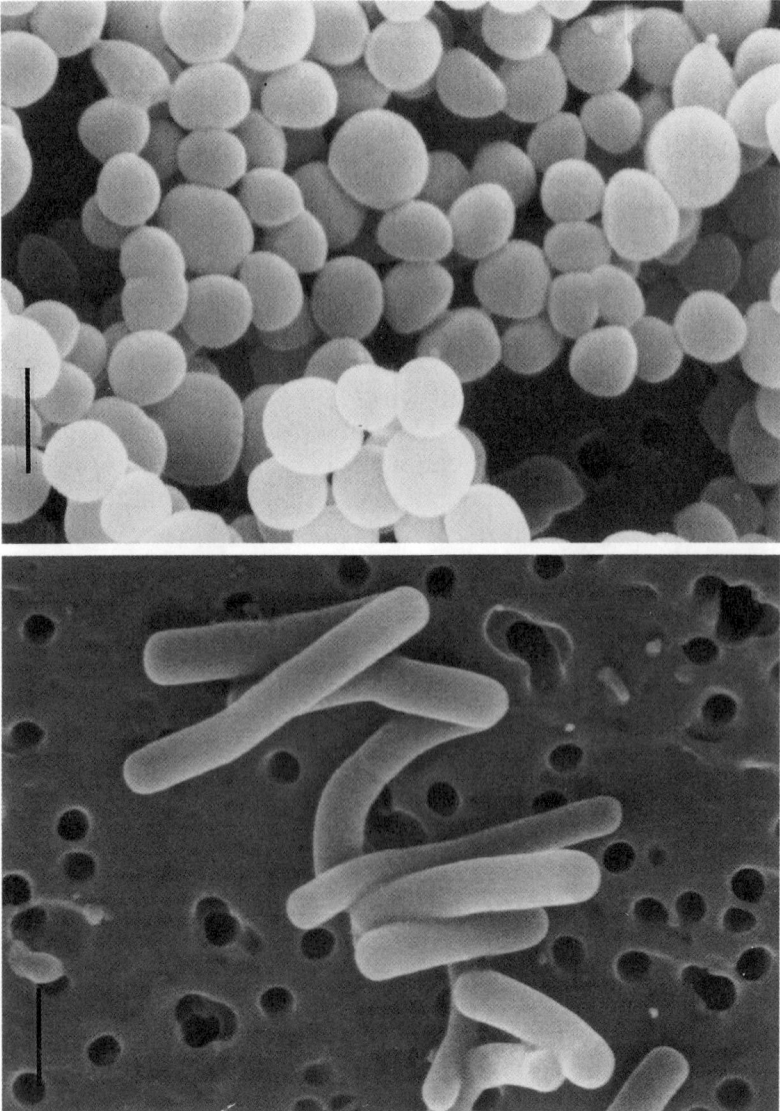

surface. A detector collects the secondary radiation and converts it into an image that can be observed on a monitor screen and photographed.

A more recent development, the TUNNELING ELECTRON MICROSCOPE scans the surface of an object with a supersharp probe and traces the object's contours with incredible precision. The movement of the probe is then converted into an image on a monitor screen. Shapes of objects as small as individual molecules have been visualized with this instrument.

Electron microscopes are useful in determining sources of particulate cleanroom contamination. For example, traces of tobacco smoke, cosmetics, or pollen can easily be identified by their characteristic appearance. Electron microscopes are enormously expensive pieces of equipment requiring highly trained operators and significant maintenance costs. It is not unusual to spend over $100,000 for such a microscope, but the volume of information that can be gained from these instruments is enormous.

Summary

Microbiology is the study of the smallest of living cells, the microorganisms. These include bacteria, yeasts, molds, algae, and protozoa. Viruses, while not considered living cells, are also considered microorganisms. Because of their small size, microorganisms cannot be seen with the unaided eye but must be observed with a microscope, either the type that uses visible light or that uses an electron beam as its source of illumination.

Most microorganisms are either beneficial to other forms of life on earth, including humans, or are neutral, but a few are known to cause spoilage and the deterioration of foods and materials and to produce disease. Microorganisms can cause problems in clean rooms through a variety of activities. Because they can conduct an electrical current, their presence in electronic circuitry could cause a short circuit and lead to a malfunction. If microorganisms or their products gain entrance into a pharmaceutical product or medical device, they could produce illness or death.

2 Growth of Microorganisms

The word growth means different things to different people. To most people growth implies increase in size as in the growth of a tree or a child. To a microbiologist, growth may also mean the increase in the size of a microbial cell, but more often it refers to an increase in the numbers of cells in a microbial population. In this chapter and for the remainder of this book, growth will usually refer to the latter meaning: an increase in the numbers of a microbial population.

Most of our knowledge of microorganisms has been gained by growing them in the laboratory, which microbiologists only learned to do a little over a hundred years ago. Growing microorganisms in the lab is a necessary and important activity for the cleanroom microbiologist. Raw materials, process water, finished products, and the air and surfaces of the manufacturing facility must be sampled for the presence of excessive numbers of microorganisms. This is necessary to confirm that all contamination control measures are operating properly. The presence of microorganisms is almost always detected by observing their growth. In this chapter we will describe some of the techniques used for growing microorganisms.

Scientific Notation

Microbial populations frequently reach enormous levels. It is not unusual, for example, to have 1,000,000,000 bacterial cells in a milliliter of growth fluid. Dealing with such large numbers can be difficult without a useful tool known as SCIENTIFIC NOTATION, which uses exponents to denote zeros or decimal places. That is, we write the number 1,000,000,000 as 1×10^9 (one times ten to the ninth power); 7200 as 7.2×10^3; and 301 as 3.01×10^2; and so on. Scientific notation will be used throughout the balance of this book to express large numbers. Incidentally, scientific notation can also be used to express very small numbers. For instance, $1/_{1000000}$ (.000001) can be written in scientific notation with a negative exponent: 1×10^{-6}. One final note on a bit of scientific jargon: The term LOG (from logarithm) is frequently used to mean a ten-fold change in something, such as a microbial population. Thus when it is said that a population is reduced by one log, it means it has been reduced by one exponent, such as from 10^7 to 10^6. The term is frequently used in connection with methods of controlling microorganisms, the subject of chapter 3.

Growth of Bacteria

Nutritional Requirements

In order to grow bacteria in the laboratory, a MEDIUM (MEDIA, plural) must be prepared. A proper bacteriological medium contains all the nutrients necessary to sustain the growth of a population of bacteria. A microbial population growing in a medium is frequently referred to as a CULTURE. Like all other living organisms, bacteria need the basic elements carbon, nitrogen, phosphorus, sulfur and other inorganic constituents, plus a source of energy. The range of nutritional requirements among the many species of bacteria is unbelievably broad. Some species can grow in a medium that contains nothing more than a few mineral salts and glucose, from which the bacteria can synthesize all of their complex nutritional needs, such as amino acids, vitamins, purines, and pyrimidines. Other species may require dozens of preformed nutrients in order to grow. The kinds of bacteria most commonly encountered in the clean room can synthesize some of their complex requirements but must have others supplied to them in the medium.

Hundreds of media recipes are available to the microbiologists, and no one medium will support even a fraction of all known bacterial species, not to mention the possible thousands of unknown species. However, because the great majority of bacteria encountered as contaminants in clean rooms generally have common nutritional needs, only one or two media are usually sufficient to grow them. The one bacteriological medium most often recommended for assessing bacteriological levels in clean rooms is one containing enzymatic digests of casein (a milk protein) and soy (bean) protein and known generically as tryptic soy medium. This medium will support the growth of many bacteria that originate from soil contamination as well as many that reside on the human body. An exception to the general use of this medium is in the culturing of bacteria that are commonly found growing in purified water systems. As mentioned in chapter 1, these organisms are capable of deriving their nutritional needs from the traces of contamination that leach from plastic piping, dust, and other sources. The concentrations of nutrients in ordinary bacteriological media are frequently inhibitory to these organisms and must be made more dilute. For example, for growing the water bacteria, tryptic soy agar is prepared at a concentration of 0.3 percent (3 gm/liter) rather than the 3 percent that is recommended for most common bacteria. A medium known as R2A was devised especially for growing purified water contaminants and consists of very low levels of nutrients.

Frequently added to microbiological media is AGAR, a polysaccharide derived from certain types of seaweed, that acts as a solidifying agent. (Microbiologists also use the word agar to refer to a medium that contains the solidifying agent plus the nutrient components.)

Before a medium can be used it must be sterilized; that is, it must be rid of all living organisms. The reason for that is clear: one does not want CONTAMINANTS (unwanted organisms) growing in the medium when trying to grow a specific organism. Methods of sterilization are covered in chapter 3. Sterile agar-containing medium is melted at 100°C and poured into sterile petri plates (Figure 2.1). On cooling, the agar will solidify when it reaches about 40°C and become a rigid, gel-like medium. When bacteria are inoculated (deposited) on the agar and if the nutritional and environmental conditions are appropriate, they will grow. When growth occurs, visible accumulations of bacterial cells will eventually form. These are called COLONIES (Figure 2.2). Each colony may contain 10^8 or more cells.

Figure 2.1. A Petri Plate containing an agar medium.

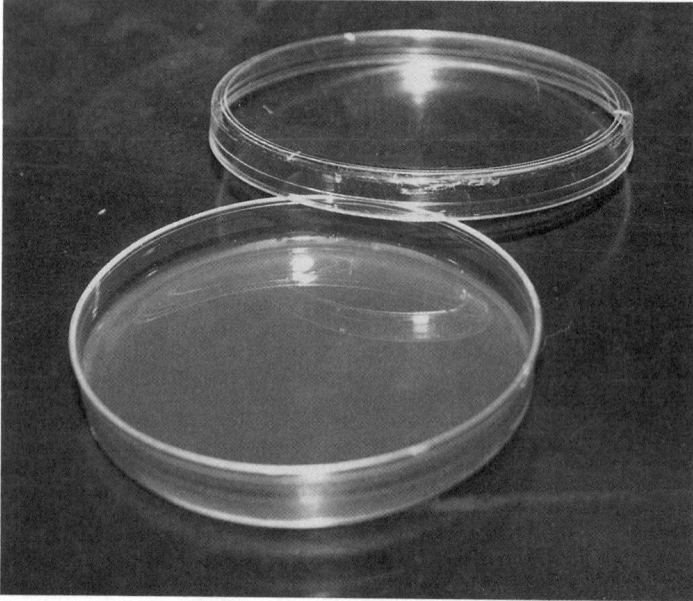

Figure 2.2. Bacterial Colonies. Accumulations of bacterial growth on an agar surface are called colonies. These colonies developed from an air sample following 48 hours' incubation.

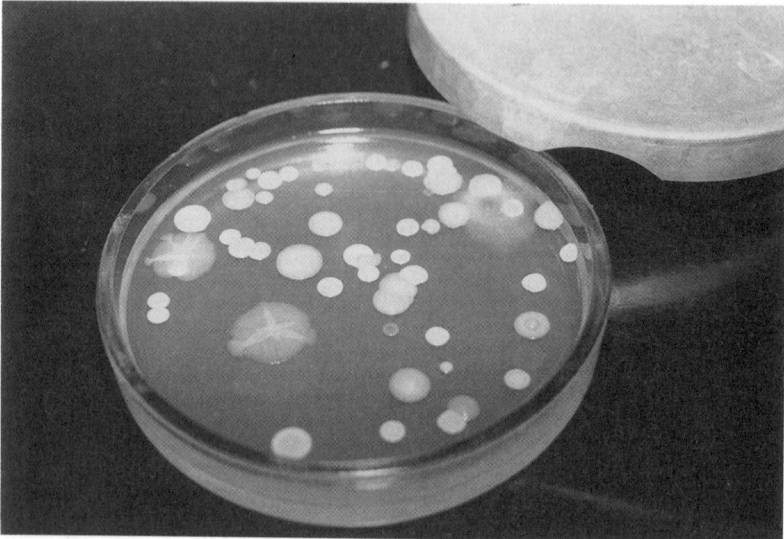

If agar is not included in the recipe, a clear liquid medium is achieved, known as BROTH. Colony formation is not possible in broth, but instead, uniformly dispersed growth normally results, creating distinct cloudiness, or turbidity. The degree of the turbidity can be used to estimate the number of cells in the culture. This procedure will be covered in greater detail later in this chapter.

In addition to supplying the nutritional needs of the microorganisms we wish to grow, certain environmental conditions also must be present. These include temperature, atmosphere, and pH.

Temperature Requirements

If one were to determine the temperatures at which growth is optimum for a variety of bacterial species, one would discover that the organisms exhibit a wide range of temperature requirements (Figure 2.3). Microorganisms are usually divided into three groups according to their optimum (best) growth temperatures:

Figure 2.3. Bacterial Growth Responses to Various Temperatures. Most species of bacteria generally fall into four groups according to preferred growth temperature: Psychrophiles grow best around 10°C; mesophiles around 35°C, thermophiles around 60°C, and extreme thermophiles, around 90°C. Note that ranges may overlap.

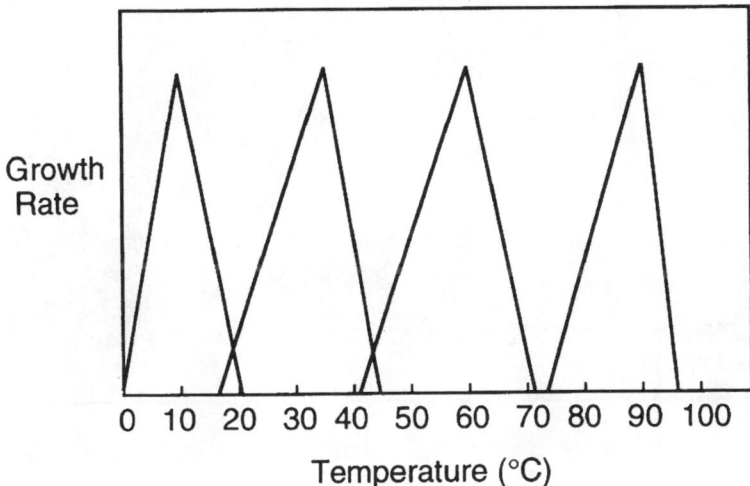

PSYCHROPHILES grow best below 20°C, some doing quite well below 5°C, which is the temperature of most refrigerator interiors; MESOPHILES find 20°C to 40°C optimum. THERMOPHILES prefer temperatures over 40°C, and some are able to grow at temperatures over 100°C, which is hotter than boiling water. To assure constant growth temperatures in the laboratory cultures are placed in an INCUBATOR (Figure 2.4), an oven-like device with a thermostatically controlled interior. Most of the bacteria encountered in the clean room are mesophiles and are usually grown at temperatures from 35°C to 37°C. An exception might be thermophilic microbes that inhabit hot water systems. These are generally incubated at 55°C.

Atmospheric Requirements

It was the great French chemist Louis Pasteur who first realized, while studying the microbiology of winemaking, that life was

Figure 2.4. A Microbiological Incubator. Thermostatically controlled, the incubator can be set to maintain the temperature appropriate for the species of microorganisms to be grown.

possible in the absence of gaseous oxygen. He showed that the wine yeasts were able to grow in wine vats where there is little or no air. Organisms that can grow in the absence of free oxygen are called ANAEROBES. Those organisms that can grow either in the presence or absence of free oxygen (such as the wine yeasts) are referred to as FACULTATIVE ANAEROBES, while those that must grow in the complete absence of free oxygen are called OBLIGATE ANAEROBES. Gaseous oxygen is actually toxic toward obligate anaerobes, and they will die if their vegetative cells are exposed to it for extended periods of time.

There are a few obligate anaerobic species of bacteria associated with soil and the human body that may occur in cleanroom contamination, some of which are of great medical significance. The bacteria that cause tetanus and botulism food poisoning are obligate anaerobes. However, the vast majority of bacteria encountered in clean rooms will be AEROBES, that is, capable of growing in a normal laboratory atmosphere. This is not to say that anaerobic contaminants should be ignored, quite the contrary. A certain proportion of all cleanroom samples should be incubated under anaerobic conditions. One may also encounter MICROAEROPHILS, bacteria that grow best in atmospheres with oxygen concentrations somewhat lower than that of ordinary air, and frequently with higher levels of carbon dioxide.

When it is necessary to grow microaerophils or anaerobes, various devices are available to control the composition of the atmosphere, including anaerobic jars and incubators (Figure 2.5) as well as special types of media, such as thioglycollate broth, which contains reducing agents that establish anaerobic conditions within a culture tube without the need for additional equipment beyond an ordinary incubator. A convenient system for culturing anaerobic bacteria involves the use of a foil pouch that contains chemicals that generate carbon dioxide and hydrogen when water is added. The pouch is placed in an airtight vessel known as an anaerobic jar. With the help of a palladium catalyst in the jar, the hydrogen that is generated combines with residual oxygen to form water. The catalyst is almost identical to the catalytic converter that is installed in automobiles to convert noxious exhaust gases to relatively harmless ones. Anaerobic conditions are soon established in the anaerobic jar as evidenced by the appearance of a strip of indicator paper that turns colorless in the absence of oxygen.

pH Requirements

pH is a chemical term that refers to the acidity or alkalinity of a liquid and covers a scale from 0 to 14, 0 being very acid and 14 being very alkaline. A pH of 7.0 is considered neutral. Table 2.1 lists the pH of several familiar liquids. Most commonly encountered bacteria grow best around neutrality, 6.5 to 7.5. Some rare species of bacteria grow optimally at pHs as low as 2.0 and can

---

Figure 2.5. Growing Anaerobic Bacteria. An anaerobic incubator (right) has provisions to adjust atmospheric composition to match organisms' requirements. An anaerobic jar (left) contains a GasPak pouch that generates hydrogen when water is added to it. A catalyst in the jar causes the hydrogen to combine with the oxygen to form water, creating anaerobic conditions.

tolerate a pH as low as 0.5. At the alkaline end of the scale, some bacteria are known to require a pH of 9.5 for best growth but can reproduce even when the pH is over 11. These bacteria that require extremes in pH (and temperature, for that matter) are usually found in outdoor habitats, such as oil wells, mines, industrial waste ponds, and alkaline or saline lakes. They are unlikely to be cleanroom contaminants except where these extreme conditions might exist, such as in hot water supplies, brine or pickling tanks, and the like. Do not assume that an environment is too harsh for microorganisms until it is checked out.

When a medium is prepared in the laboratory, the pH should always be tested to assure that it is within the optimum range for the desired organisms and adjusted with acid or base when necessary. During the growth of the bacteria, the pH probably will not remain constant but will shift as a result of the organisms' metabolism. It is not unusual for a culture to produce so much acid as a waste product that the pH soon shifts below the organism's

TABLE 2.1.
pH OF SOME FAMILIAR LIQUIDS

Liquid	pH
1 N hydrochloric acid	0.1
Human gastric juice	1.0–3.0
Lime juice	1.8–2.0
Vinegar	2.4–3.4
Wine	2.8–3.8
Human urine	5.0–8.0
Human saliva	6.5–7.5
Pure water	7.0
Human blood	7.3–7.5
Fresh egg	7.6–8.0
0.1 N sodium bicarbonate	8.4
Drain cleaner	14

pH range and growth stops. To avoid this problem, one can either periodically check the pH of the medium and adjust it with sterile base, or incorporate a BUFFER in the medium. A buffer is a substance that can counteract shifts in pH.

Checking the pH on a continuous basis is usually not practical in ordinary laboratory cultures although apparatuses for growing moderately large volumes of microorganisms, called fermenters, have the capability of automatically monitoring and adjusting the pH. The practice is also common in large industrial microbial processes. The addition of buffers to the medium is a convenient and inexpensive means of preventing pH shifts in small laboratory cultures, but they probably are not practical on large industrial scales. Many constituents of common bacteriological media, such as peptones, are natural buffers and are frequently adequate for most purposes. If a medium does not contain these natural buffers or if greater buffering capacity is required, a number of chemical buffers can be used, but they must be chosen carefully; some are toxic to microorganisms and were not intended to be used in growth media.

Binary Fission and the Bacterial Growth Curve

Bacteria grow, that is, increase their numbers, by BINARY FISSION, which means that one cell divides into two, which in turn becomes four, then eight, sixteen, and so forth (Figure 2.6). The growth of a bacterial population can thus be expressed mathematically in terms of powers of 2:

$$\text{final population } P = 2^n, \tag{1}$$

where n is the number of times the population doubles. Whenever a population doubles, it is said to have gone through one GENERATION. For example, one cell becomes 4, 32, 1024, and 1,048,576 cells after 2, 5, 10, and 20 generations, respectively. If we start with more than one cell (which is almost always the case), the equation becomes

$$P = P_0 \times 2^n, \tag{2}$$

where P_0 is the starting population.

When we graph the viable population of a growing culture of bacteria against time, a curve such as is shown in Figure 2.7A would result. (The VIABLE population is defined as those cells that are capable of growing when transferred to another medium.) Notice that because the population increases so rapidly, we soon run out of space on the vertical (population) axis. A more manageable curve is obtained (Figure 2.7B) by plotting the logarithm of the population versus time. Now one sees that the GROWTH CURVE, as it is called, is actually a straight line, which is to be expected, because the increase in population follows the exponential term 2^n. Whenever a bacterial culture ex-

Figure 2.6. Bacteria Undergoing Binary Fission. Thin sections of *Staphylococcus* cells are seen in various stages of division. Some (A) are just beginning division, while others (B) appear to have completed cross-wall formation and will soon separate into two daughter cells. Some cross-walls appear to be off center (C) because cells were cut at random angles. The cells were stained with uranium. Bar = 1 μm. (Electron micrograph courtesy of Tai Wu, California State University, Long Beach.)

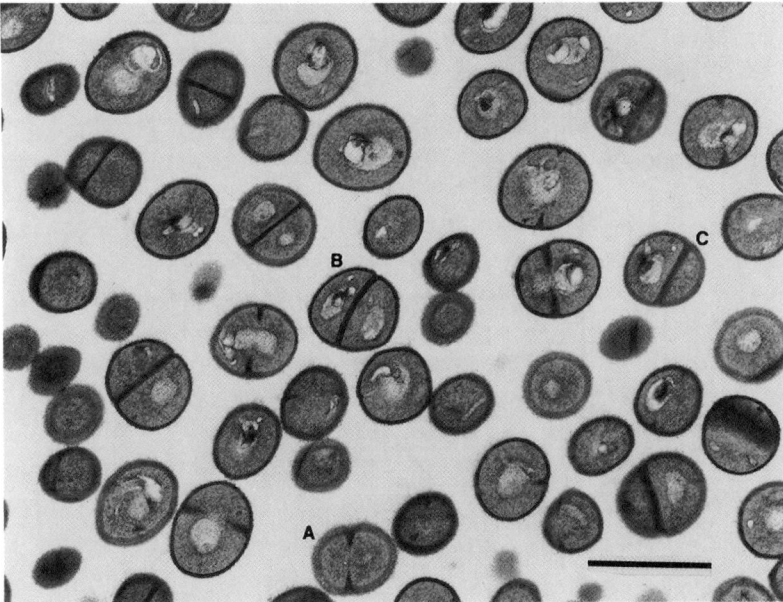

Figure 2.7. Growth Curves of Bacteria. A. When population is plotted on a linear scale against time, the population increases so rapidly that the curve runs off the chart very quickly. B. When the population is plotted on a logarithmic scale, the enormous population increases characteristic of bacteria can be easily accommodated. Also, the curve becomes a straight line.

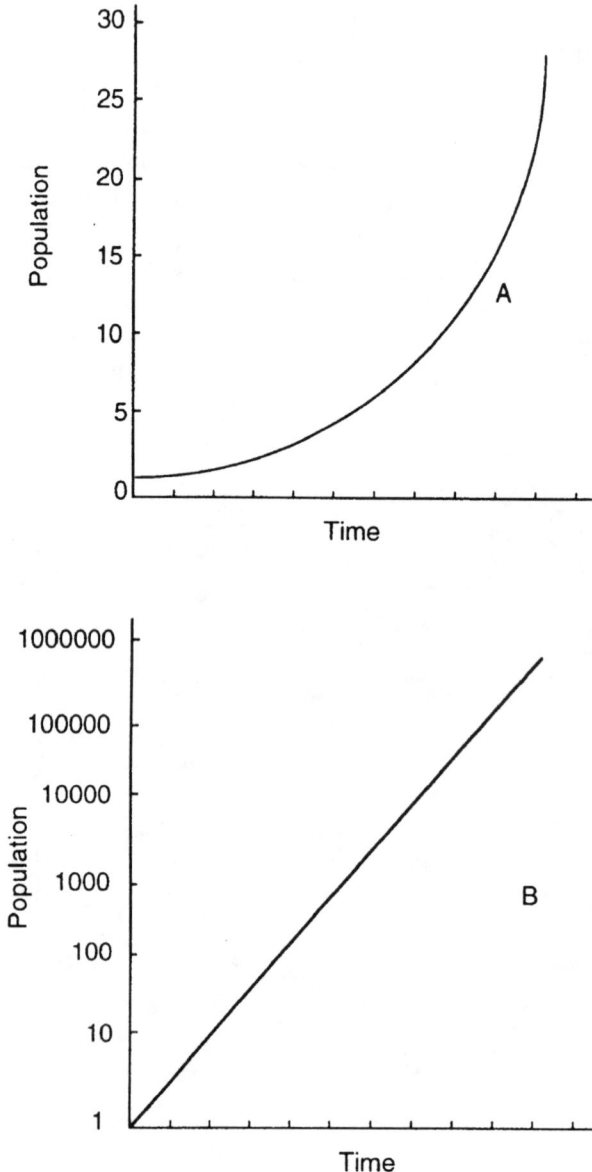

hibits a linear growth curve as in Figure 2.7B, it is said to be in the LOGARITHMIC or LOG PHASE of growth. Three other principal phases exhibited by bacterial cultures are LAG, STATIONARY, and DEATH phases (Figure 2.8).

Lag Phase

When a medium is first inoculated with one or more bacterial cells, growth normally does not begin at once. This delay is called the LAG PHASE, and its length depends on the nature of the medium from which the cells were transferred and the new medium. Nevertheless, the cells are physiologically very active during the lag phase, retooling their metabolism to match the conditions of the new medium. When all necessary enzymes and cofactors have reached optimum levels, growth begins.

Log Phase

Soon most, if not all, the cells in the new medium begin to divide at a fairly constant rate. Division is not synchronized, however,

Figure 2.8. A Complete Growth Curve. A. The lag phase during which the cells, while not dividing, are very active metabolically getting ready for the next phase. B. The log phase when cells are dividing at their maximum rate. C. The stationary phase when net population increase is zero. D. The death phase when more cells are dying than are dividing.

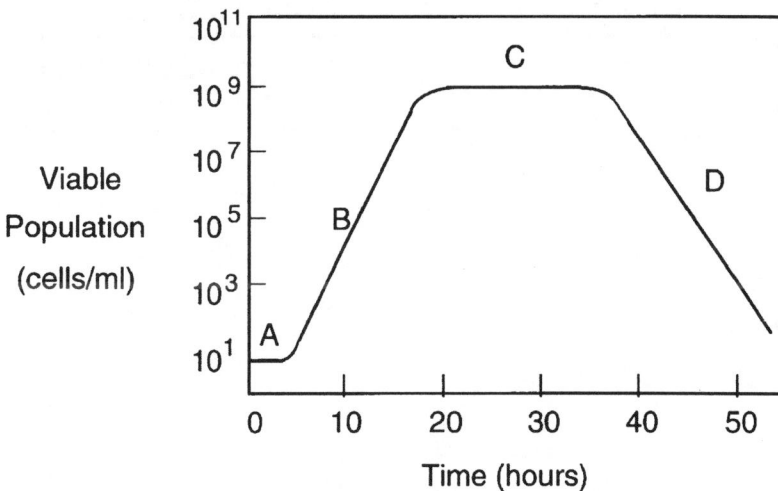

for the cells are dividing quite randomly throughout the culture. That is, cells are beginning division at different times, resulting in the smooth, straight-line characteristic of the LOG PHASE.

During the log phase, the cells achieve BALANCED GROWTH, where every constituent and property of the cells, proteins, nucleic acids, cell wall, total mass, and so forth, are increasing at the same rate as every other constituent. The GENERATION TIME, that is, the interval of time a given cell population needs to double in numbers, is also constant during the log phase. However, within a population of bacterial cells, generation times probably vary somewhat from cell to cell. The mean generation time of a culture can be calculated by rearranging equation (2) and bringing in t, the elapsed time the cells were growing:

$$G = \frac{t}{n} = \frac{t}{3.3 \, (\log P - \log P_0)} \tag{3}$$

where G = generation time in minutes, P_0 is the starting population, and P is the population after t minutes (3.3 is $1/\log 2$). Typical generation times of some familiar bacteria are shown in Table 2.2. They span from 11 minutes for *Bacillus stearothermophilus* to 2000 minutes (over 33 hours) for *Treponema pallidum*. These generation times were determined under optimal growth

TABLE 2.2.
GENERATION TIMES OF SOME BACTERIA

Species	Generation Time (minutes)
Bacillus stearothermophilus	11
Escherichia coli	20
Streptococcus lactis	30
Lactobacillus acidophilus	75
Mycobacterium tuberculosis	360
Treponema pallidum	2000

conditions. Generation times may vary depending on the medium and other growth conditions, such as temperature.

The log phase does not continue indefinitely. The depletion of necessary growth factors and the accumulation of toxic waste products eventually result in a relatively rapid flattening out of the growth curve. The culture is now entering the stationary phase.

Stationary Phase

As the log phase progresses, the nature of the medium changes. Many essential growth factors become depleted or nearly so, and the pH probably shifts out of the optimum range for the bacteria. The cells' metabolism becomes increasingly inhibited by their own toxic waste products, including organic acids, alcohols, peroxides, carbon dioxide, and the like. During the STATIONARY PHASE, there is no apparent increase in the viable population either because cell division ceases entirely, or because equal numbers of cells are dividing and dying.

Depending on the species of bacteria and the nature of the medium, the stationary phase may last a few minutes, or perhaps several hours, during which time the viable population appears to remain level.

Death Phase

Eventually the viable population of the bacterial culture begins to drop exponentially in an almost mirror image of the log phase, but probably with a different slope. Certain species exhibit a very steep DEATH PHASE so that few, if any, viable cells are recoverable after 48 hours or so, while with other species viable cells may be detected even after a year or more.

The Chemostat

Even with strong buffering, the life of a bacterial culture is limited as described above. For most bacteria the duration of the log phase is usually not more than a few hours. There is, however, an apparatus that makes it possible to maintain a bacterial culture in the log phase essentially indefinitely. The apparatus is called a CHEMOSTAT and is simply a growth vessel into which a slow but constant supply of fresh medium is added and from which excess medium and cells are allowed to overflow. The fresh medium contains one limiting nutrient that acts as a throttle to control the growth rate of the bacteria. Thus, growth rate can be adjusted by merely regulating the rate at which the fresh medium is pumped into the vessel. The chemostat has a number of research and applied uses, such as being able to have a ready supply of log phase bacteria at any time ("bacteria on tap").

Determining Bacterial Populations in Cultures

The creation of a bacterial growth curve requires measuring the cell population at frequent intervals during the life of the culture. We can determine the viable population by conducting a plate count or estimate the total cell population by using one of a number of physical or chemical assays. The distinction must be made between the viable cell population and the total cell population because a certain proportion of the cells of every bacterial population will be nonviable. As you recall, viable means capable of growth if transferred to an appropriate medium. The exact proportion of nonviable cells varies with the age of the culture and other conditions, but during the log phase it is usually not more than a few percent. During the stationary and death phases, however, the proportion of nonviable cells becomes significant.

Viable Plate Count

To assess the viable population, carefully measured samples of the culture are removed, diluted a suitable number of times, and plated onto an appropriate medium (Figure 2.9). Following the incubation of the medium under conditions that would support the cells that are to be assayed, colonies are counted and the results are related back to the original suspension by the following calculation:

$$\text{Colony-Forming Units/ml} = \frac{\text{number of colonies}}{\text{dilution factor} \times \text{plating volume}}$$

Notice that the results of the population determination are in terms of colony-forming units per milliliter (CFU/ml). This

Figure 2.9. The Viable Assay. A bacterial cell suspension is first accurately diluted in several sequential steps (top). Samples of suitable dilutions are then spread onto appropriate agar media with a sterile glass rod (bottom). The plates are incubated and the colonies that form are counted. From these results the viable population of the original suspension is calculated.

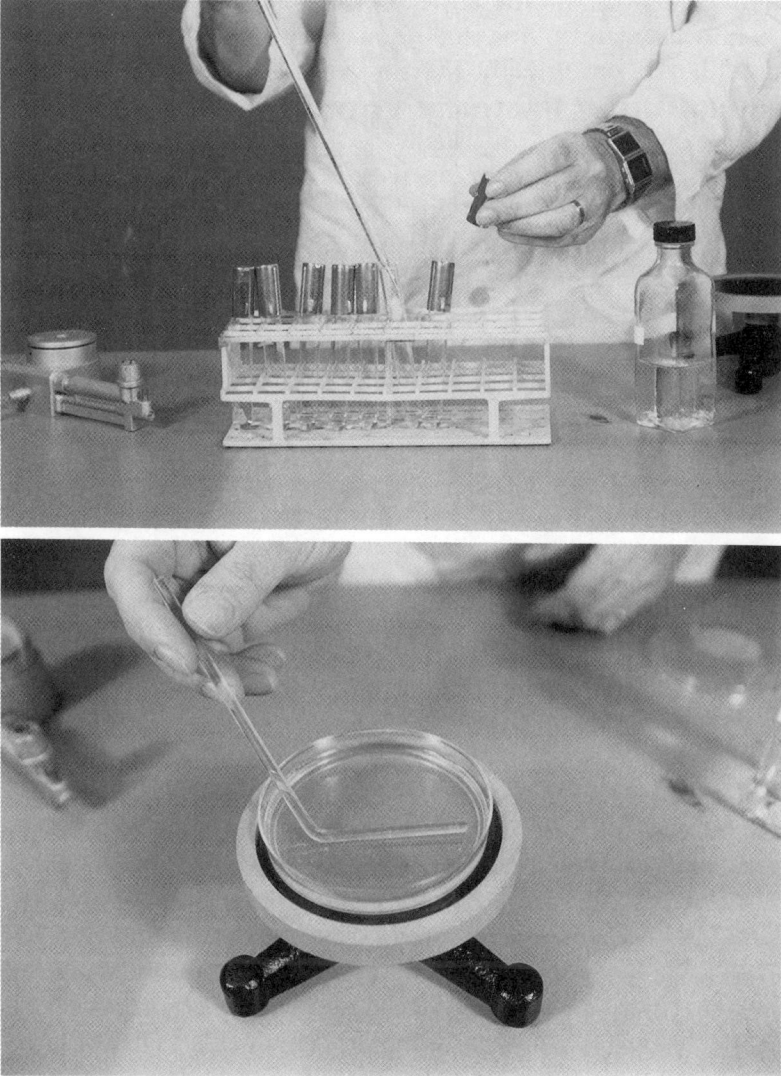

awkward expression is necessary because each colony that appears on an assay plate may be the result of the growth of a single cell or it may have originated from a clump of cells that either failed to separate on division or aggregated at some later time. Because there is no way to distinguish between these alternatives, expressing the population in colony-forming units per milliliter recognizes this uncertainty and avoids any implications as to the true origin of the colonies. Obviously then, viable counts are only estimates of the true population of a culture, but if done carefully and consistently, they are very reliable. The terms *dilution factor* and *plating volume* are explained below.

As noted earlier, it is usually necessary to make suitable dilutions of the sample to be assayed because many bacterial suspensions, such as a typical broth culture, may contain upward of 10^8 cells per milliliter. As a general rule, the ideal plate count is considered one in which not more than 300 nor less than 30 colonies appear on the plates. Assuming a population of 10^8 cells/ml and a plating volume of 0.1 ml, the final dilution tube should have between 300 and 3000 cells/mL, requiring a 100,000-fold (10^{-5}) dilution of the original suspension.

The lower limit of the 30–300 colonies rule is based on statistical considerations. The standard error associated with plates with fewer than about 30 colonies becomes a significant and unacceptable fraction of the count, nearly 20 percent. However, from a practical standpoint very low colony counts are frequently seen in environmental samples taken in the more stringent cleanroom classes. This problem is discussed in chapter 5.

As colony counts exceed about 300 per standard size petri plate, the probability becomes significant that two or more colonies will overlap and erroneously be counted as one. This frequently becomes a problem when assessing microbial loads of raw materials, where microbial levels may be very high. In this case, dilutions must be made of the samples.

Dilutions are carried out in sterile blanks prepared with saline, phosphate buffer, water, or broth. For tenfold dilutions, 4.5 ml blanks are recommended to which is added 0.5 ml aliquots. Hundredfold dilutions are best handled by adding 0.1 ml aliquots to 9.9 ml dilution blanks. The individual dilution factors for each step are then multiplied to arrive at the final dilution factor that is used to calculate the population.

A fresh pipette should always be used for each dilution step, for bacterial cells have a tendency to adsorb to the inner walls of pipettes and may be released two or more steps later resulting in a large error in the assay results.

The plating volume is the volume that is transferred to the agar. Plating volumes are generally 0.1, 0.5, or occasionally 1.0 ml. Volumes larger than 1 ml are usually not recommended because of the limited amount of fluid the hardened agar can absorb if using the so-called SPREAD PLATE method. In this procedure diluted aliquots are spread uniformly over the surface of the agar with a sterile glass rod that is bent in the shape of a hockey stick (Figure 2.9). Batches of rods may be sterilized in advance and used once, or one rod may be repeatedly used and sterilized by dipping in alcohol and igniting the alcohol with a gas flame. The rod should not be overheated in the gas flame; it is only necessary to ignite the alcohol to sterilize the rod.

An alternate assay method, the POUR PLATE, involves transferring the plating volume to a tube of melted agar (held not higher than 45°C) that is thoroughly mixed by spinning between the palms of the hands. The contents of the tube are then poured into a sterile petri plate and allowed to solidify. This technique results in submerged as well as surface colonies that are counted, and the population is then calculated in the same manner as was done with the spread plates. Alternatively, the plating volume may be delivered directly to an empty plate followed by the molten agar. The plate is then gently swirled to mix the two components.

Total Cell Count

Total cell counts enumerate all cells, live or dead. One rapid, simple technique for determining total bacterial populations involves the use of a microscope and a Petroff-Hausser counting chamber (Figure 2.10). The chamber consists of a microscope slide on which is engraved a pattern of squares. Each square is 0.05 mm (millimeters) on a side. Ridges 0.02 millimeters high support a cover slip above the grid. Thus each square as observed in the microscope can be thought of as a box $0.05 \times 0.05 \times 0.02$ mm, or 5×10^{-8} milliliters.

A small volume of bacterial suspension is placed between the cover slip and the grid, a count is made with the aid of a microscope, and the mean number of cells per square is calculated. This value when multiplied by the factor 2×10^7 results in the total number of cells per milliliter in the original suspension.

Because it is not possible to distinguish live cells from dead cells in the Petroff-Hausser chamber, this method has obvious limitations, such as the inability to follow accurately the stationary and death phases of the growth curve. Also, the method is not applicable to population densities much below about

Figure 2.10. Petroff-Hausser Counting Chamber. Engraved slide in its holder (top), and grid (bottom) as it appears in the microscope. Each of the smallest squares in the grid is 50 μm on a side.

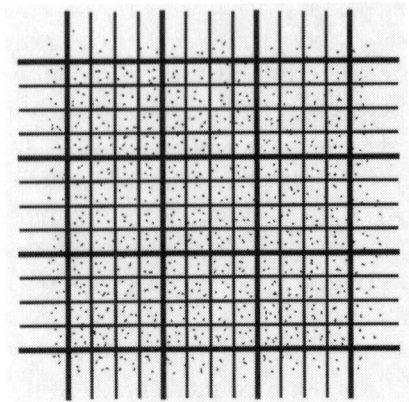

10^7 cells/ml (less than one cell per square) although a population estimate may be made by applying the Poisson Distribution to the count:

$$\text{cells/ml} = -ln(P_0) \times 2 \times 10^7$$

P_0 = the proportion of squares that contains no cells, that is:

$$\frac{\text{number of squares with no cells}}{\text{number of squares counted}}$$

Instrumental Methods

A number of automatic and semiautomatic instrumental methods are available for assessing microbial populations. They all are rapid and convenient, many capable of returning results in a few minutes or less. Some of them, however, suffer from one potentially serious shortcoming— they are not very sensitive, requiring microbial populations in liquids of at least 10^6–10^7 cells/ml. In addition, most of these methods are unable to distinguish live cells from dead cells and are, thus, incapable of determining viable populations. They still perform a number of useful and often indispensable functions in the microbiology laboratory.

Turbidometry and nephelometry

As pointed out earlier, bacterial cells are nearly as small as the wavelengths of visible light and they are colorless and transparent. Because of their small size and transparency, bacteria scatter light rather than absorb it. Light scattering is defined as the redirection of a beam of light as it passes through a small, transparent object (Figure 2.11). The intensity of the light is not reduced;

Figure 2.11. Light Scattering by a Spherical Bacterial Cell. Because the cell is very nearly the size of the wavelength of the light beam, its direction is slightly altered, but its intensity is the same.

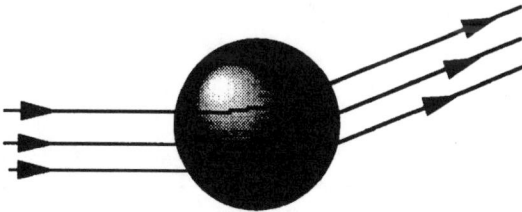

the light's path merely changes direction. Clouds are visible for the same reason. They are composed of small, transparent water droplets that should be invisible, but because of light scattering, they stand out quite noticeably.

A broth culture of bacteria appears turbid to us because of the light scattering by the cells. The angle of scattering is seldom more than a few degrees but is sufficient to notice by eye or measure by appropriate instruments. Because the light beam is diverted from its normal path, its intensity appears to be reduced when observed by a light measuring device that would be placed in the path of the normal beam as what one might do in an ordinary spectrophotometer. This technique is known as TURBIDOMETRY. By preparing a standard curve ahead of time in which optical densities are plotted versus known populations, an estimate of the bacterial population of a sample can be obtained (Figure 2.12).

Figure 2.12. Turbidity Standard Curve. To determine microbial populations by turbidity a curve such as this would be prepared using suspensions of known populations. The turbidity of an unknown microbial suspension is then measured and with the aid of the curve its population can be read on the vertical axis.

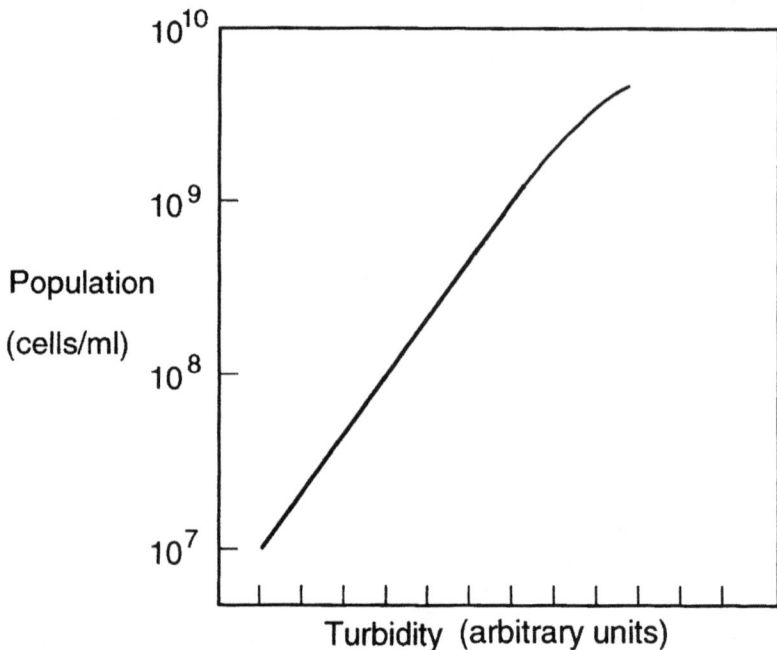

The accuracy of methods based on light scattering is highly dependent on the care taken in preparing the standard curve. Light scattering is a function of the size and shape of the particle, which means the standard curve must be prepared with cells that are dimensionally the same as those that are to be measured. Preferably the same species should be used if known. Wavelength is also an important factor: the smaller the wavelength the more intense the scattering and the more sensitive the determination. The shortest wavelength that is practical is around 400 nm, provided the cells are suspended in buffer or saline. Broth contains materials that absorb strongly at these short wavelengths and therefore 530 to 600 nm is recommended for measuring turbidities of broth cultures.

One can measure the intensity of the deviated light with the use of a special instrument called a nephelometer where the light detector is placed slightly off-axis from the light beam, rather than on-axis as in a spectrophotometer. NEPHELOMETRY has the advantage of greater sensitivity where populations down to about 10^6 cells/ml can be measured routinely.

The use of light scattering to estimate bacterial populations, particularly turbidometry, goes back nearly to the turn of the century, and in spite of its shortcomings, it continues to be one of the most popular methods for that purpose.

Electrical resistance and impedance

Microbial cells are relatively poor conductors of electricity and, consequently, if one or more cells are placed in the path of an electric current, the degree of attenuation of the current can be a measure of the population present. In instruments such as the COULTER COUNTER (Coulter Corp., Miami, FL), as one cell at a time passes through the current path it triggers the counter. Therefore, the population can be estimated with considerable sensitivity and accuracy, provided background counts due to dust and cell debris are not excessive. Less sensitive methods are based on measuring the electrical impedance of an entire suspension of bacteria. Impedance is the attenuation of an alternating (A.C.) electrical current. Not only do the bacterial cells affect the impedance of the suspension, but their metabolic products also do. This method has been applied to automated instruments in the clinical microbiology laboratory for the assessment of antimicrobial drug susceptibilities.

Radiometric Methods

Carbon dioxide is a universal by-product of aerobic microbial metabolism. By feeding microorganisms a variety of radioactive ^{14}C-labeled nutrients, nearly all bacteria will release much of the radioactivity in the form of $^{14}CO_2$, and under standard conditions, the amount of radioactivity released is proportional to the bacterial population. Cultures are maintained in sealed bottles, and periodically a sample of the headspace is removed and measured for radioactive gas. The obvious advantage of this system is that only metabolizing cells will produce CO_2, and hence, one can get a good estimate of the viable population.

ATP Analysis

ATP (adenosine triphosphate) is an important constituent of all living cells. It plays a number of roles, including a precursor in DNA synthesis and an energy transport molecule. The bond that connects the third phosphorus to the adenosine is referred to as a high energy bond, which means that when the bond is broken a considerable amount of chemical energy is released to supply energy to other reactions in the cell. The amount of ATP in a viable cell is generally constant, so a measure of the ATP content of a bacterial culture is a good measure of the number of viable cells present. When a cell dies, its ATP is usually quickly depleted; consequently, only viable cells have significant amounts of ATP.

A very sensitive method of ATP analysis has been developed that is based on the light-emitting reaction of luciferin, the molecule that is responsible for the light emission in fireflies. ATP is extracted from the bacterial culture to be enumerated. In the presence of the enzyme luciferase the luciferin emits one photon for each molecule of ATP present. A sensitive photometer is used to detect the amount of light emitted by the luciferin/luciferase reaction, which is proportional to the number of viable cells present in the sample. With the help of properly prepared standards, one can then get an estimate of the viable bacterial population. The lower limit of sensitivity for this method is about 10^4 cells/ml.

Epifluorescent microscopy

Another technique for counting bacteria in liquid suspensions is by trapping them on a membrane filter. The filter can be placed on a nutrient agar surface to allow the formation of colonies, which can be counted as in a viable count. Alternatively, the bacterial cells on the filter can be counted directly with a microscope. Normally, this approach offers considerable difficulties, for much of the particulate matter that is trapped by the filter is nonmicrobial. The trouble lies in the difficulty in distinguishing microbial cells from the noncellular trash. There are, however, fluorescent dyes, such as acridine orange, that preferentially stain microbial cells. An epifluorescent microscope is used to examine filters prepared by this technique. An ultraviolet light source is directed down through the objective lens. Viewed in a darkened room, stained bacterial cells show up as fluorescent objects against a dark background whereas noncellular debris is either dimly outlined or invisible.

Chemical Methods

By measuring principal chemical components of the cells, like DNA, protein, total nitrogen, or dry weight, a good estimate of the population size can be made, but again, attention must be paid to the sensitivity limitations of the method. For example, the popular Lowry reagent can measure protein quantities down to about 50 mg. This is equivalent to about 5×10^8 cells. Recognition must again be made of the fact that, except for ATP measurements, chemical determinations are measuring total cell populations without regard to viability.

It is quite clear that no one technique to assess bacterial populations is ideal. Counts based on turbidity, microscopic counts, or electronic counts are nearly instantaneous, but they fail to distinguish live cells from dead ones and most lack sufficient sensitivity to detect low numbers of organisms. Chemical methods are time consuming and frequently lack sensitivity. The viable plate count produces accurate and sensitive viable population estimations but with a delay of at least 12 to 18 hours. Many laboratories use a combination of two or more of these methods, using when possible an instantaneous procedure as a guide and the viable plate count data for final calculations.

Counting Fungi

Mention was made in chapter 1 that fungi can be serious microbial contaminants in certain types of products, requiring the environmental monitoring of fungal populations as well as bacteria. It should be remembered that molds do not replicate like bacteria, by binary fission, but form filamentous mycelia that eventually produce thousands of spores. Thus, the determination of mold populations is not nearly as clean and precise as the determination of bacterial populations, and it must be considered only semiquantitative at best. The assessment of yeast populations is more reliable and is akin to determining bacterial populations.

While fungi will grow on most common bacteriological media, there are several reasons to use media specifically devised for fungi. Fungi generally grow more slowly than bacteria, frequently requiring several days to produce visible colonies, whereas bacterial colonies usually reach maximum size in less than 24–48 hours. In dealing with environmental samples bacteria will almost always be present with the fungi. When trying to grow fungi, it is therefore useful to avoid bacterial growth in order to prevent overgrowth and competition for nutrients. There are several media that encourage fungal growth while inhibiting the growth of most bacteria. Such media normally have higher levels of sugar and are adjusted to lower (more acid) pHs, both conditions being generally inhibitory to bacteria. In special cases, antibiotics that have little effect on the fungi can be added to the medium to inhibit the bacteria.

One of the most popular media for growing the more common fungi is Sabouraud Dextrose agar. Its pH is adjusted to 5.6, and it contains 4 percent glucose, at least four times the concentration normally used in bacteriological media. Mycological Agar and Malt Extract Agar are also used for culturing environmental fungi. In addition fungi normally prefer incubation temperatures around 22°C to 25°C, which is below the optimum temperature for many common mesophilic bacterial cleanroom contaminants.

Counting Viruses

Viruses must grow by synthesis and assembly within a host cell. Therefore, to grow viruses in the laboratory appropriate host cells must be found and techniques for their growth and maintenance must be developed. The development of the techniques of CELL

CULTURE, the ability to grow cells of higher plants and animals in laboratory glassware, has made it possible to cultivate many types of viruses. Human cells, for example, can be grown in bottles and infected with any of a number of human or other animal viruses. As one might expect, the nutritional and atmospheric requirements necessary to grow animal cells are considerably more complex, and the skills involved are more demanding than those used in the cultivation of most other microorganisms.

Not all viruses can be grown in cell cultures, and microbiologists sometimes have no choice but to use a whole animal host in order to obtain viruses to study. Alternatively, fertile chicken eggs have proven to be susceptible to a number of animal viruses and can act as hosts for virus cultivation. Some viruses for commercial vaccine production are produced in chicken eggs.

Once a method for growing a particular virus in the laboratory has been developed, the next step is to measure the actual extent of virus multiplication by carrying out an assay of the virions that have been produced by the host cells. This number is frequently referred to as the TITER. Viruses do not form colonies, but for many viruses an assay method somewhat akin to a colony count may be applied. Called a PLAQUE ASSAY, the method takes advantage of the effect some viruses have on their host cells. Many viruses produce a CYTOPATHIC EFFECT, visible damage to their host cells as a result of their presence. The most severe effect is the death and lysis of the host. If viruses are made to infect a one cell-thick layer of host cells that have been growing on a solid surface such as agar or glass, (a MONOLAYER), the localized destruction of host cells will produce visible clearings, or PLAQUES in the layer of cells. The plaques can then be counted and the titer calculated in a manner identical for determining bacterial populations by colony counts.

Viruses are generally not a problem in most clean rooms involved in the manufacture of pharmaceuticals and medical devices. A significant exception involves those products that are derived from cell cultures, such as certain vaccines and products derived from genetically engineered animal cells, which could be a source of viruses. Additionally, such operations must be protected from viruses that could infect the cells. For those facilities involved with these activities, microbiological monitoring of raw materials, finished products, and the environment must include viruses.

Aseptic Technique

It is clear that when one is involved in growing a culture of microorganisms for any reason, it is imperative that no unwanted microorganisms be allowed to enter the culture for the entire process from the preparation of the sterile medium and its initial inoculation to the end of an experiment or production process. The precautions that are practiced by the microbiologist and other facility personnel to ensure that the operations are not microbially contaminated are generally referred to as ASEPTIC TECHNIQUES. Some basic aseptic techniques are illustrated in Figure 2.13. Aseptic techniques must become second nature to be

Figure 2.13. Aseptic Technique. Flaming the mouth of a vessel immediately before and after its inoculation (left) and the flaming of an inoculating loop (right) before and after its use are two examples of methods of avoiding contamination.

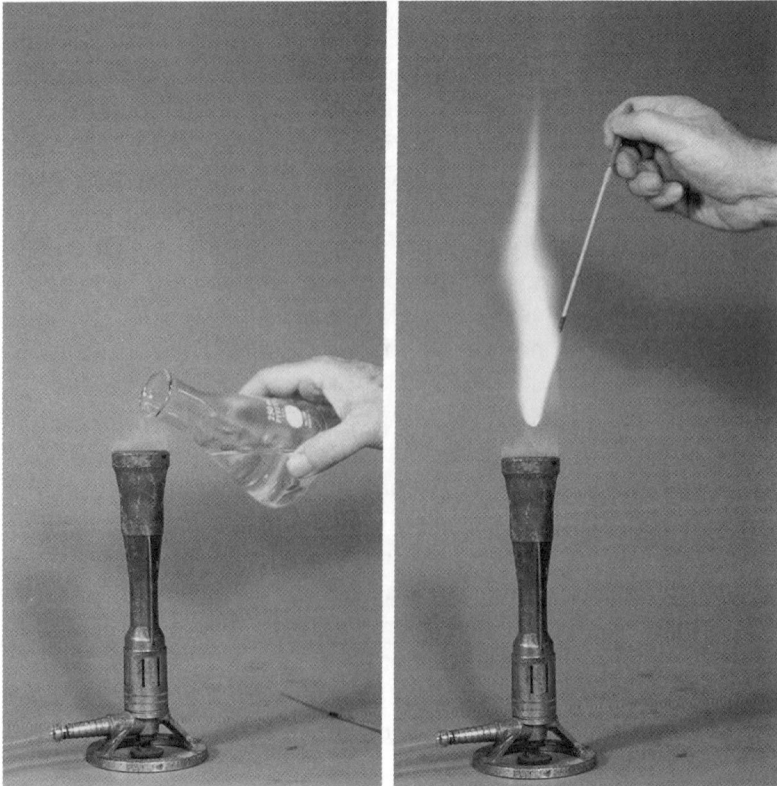

practiced the entire work day, for a contaminated culture is useless and, in some situations, dangerous. All media are sterilized prior to use, and the mouth of any vessel that is to be inoculated is flamed before and after inoculation as is the inoculating loop or needle. In addition, all used materials and glassware that are contaminated must be sterilized prior to reuse or discarding. In the manufacturing facility, similar precautions are carried out. Many of these precautions are discussed in chapter 4.

Summary

Many microorganisms, such as bacteria and fungi, can be cultured in the laboratory on various nonliving media that contain all the necessary nutrients. In the case of viruses and other more fastidious microorganisms, living cells or whole plants or animals must be used. In either case, care must be taken to supply all the chemical and physical (temperature, atmosphere, etc.) conditions necessary to support the desired organisms and when appropriate their hosts as well.

There are many techniques at the disposal of the microbiologist for measuring microbial populations. Some of these techniques determine only viable populations, others measure total (living and dead) cell numbers. There is no one ideal method, and frequently one must rely on two or more different techniques to satisfy a particular need.

Regardless of the method used to grow microorganisms, the avoidance of contamination must be a primary goal. This is accomplished by careful adherence to aseptic techniques.

3 Methods of Controlling Growth and Activities of Microorganisms

Because of the undesirable activities of some microorganisms, such as their involvement in food and drug spoilage, deterioration of materials, infectious disease, and interruption of electronic microcircuitry, it is frequently necessary to control their growth or activities by various physical or chemical means. Heat, cold, or filtration are examples of physical methods, whereas the use of disinfectants and antiseptics represents chemical means. In this chapter we will explore these methods, but first we must deal with some important definitions.

DISINFECTION refers to the reduction in numbers of the vegetative form of pathogenic microorganisms present in or on an inanimate object. PATHOGENIC means causing disease. A DISINFECTANT is an agent that is used in disinfection. Bacterial spores are usually not affected by most common disinfectants.

ANTISEPTICS are chemicals that kill or inhibit pathogenic microorganisms and are safe to apply to the body. Like disinfectants, antiseptics have little effect on bacterial spores.

DECONTAMINATION refers to the reduction of the microbial population in or on an object to some lower value but not necessarily to zero.

SANITIZATION is the reduction of microbial populations to levels considered safe by public health standards. Restaurants and cocktail bars frequently use SANITIZERS for a final rinse of serving utensils and food handling equipment. This term is occasionally used inappropriately to describe disinfectants used in clean rooms and other industrial applications.

STERILIZATION means the complete removal or destruction of viable organisms. Sterility is an absolute condition; an object is either sterile or it is not whereas an object that has undergone disinfection or decontamination may still be contaminated. Because sterilization requires much harsher conditions than decontamination, it is often useful to recognize instances when decontamination would suffice, which would avoid conditions that might be damaging to the object to be treated.

VIABLE means the ability of an organism to reproduce when placed in a suitable environment. Conversely, NONVIABLE is defined as the lack of a cell's ability to reproduce when it is placed in an environment that would normally support its growth. Occasionally microbial cells may appear nonviable under one set of conditions and viable under another set. It is therefore important when reporting the results of the testing of a disinfectant or a sterilization process, or when conducting environmental sampling, to define the type of medium, the incubation temperature, and other culture conditions.

The BIOBURDEN is the viable microbial population in or on an object just before it is sterilized.

Physical Methods of Control

Heat

Two forms of heat are used in decontamination and sterilization: wet or moist heat and dry heat. Moist heat refers to situations where the object to be sterilized is in contact with saturated water vapor, that is, at 100 percent relative humidity (RH). Dry heat means anything less than 100 percent RH. The two kinds of heat are discussed separately because one is considerably more efficient than the other. Moderate numbers of bacterial spores are reliably killed by moist heat in 20 minutes at 121°C whereas dry heat would require over six hours at the same temperature to kill an identical spore suspension. This dramatic difference is due to the mechanism of killing of the two types of heat. Moist heat kills by rapidly coagulating cellular protein, a reaction much the same as when we hard-boil an egg. Dry heat sterilization kills by the slower processes of oxidation and desiccation.

Under ideal circumstances a population of microorganisms exposed to a high temperature dies off in an exponential fashion.

That means that a graph plotting the logarithm of the surviving population versus time of exposure results in a straight line (Figure 3.1). The slope of the line reflects the rate of death of that particular organism, that is, its relative resistance to heat. Thermal death is a complex phenomenon that is dependent on a number of factors, some of which are listed in Table 3.1. Killing curves

Figure 3.1. Bacterial Heat Killing Curve. An ideal killing curve for a population of 10^8 bacteria held at 160°C. The viable population is reduced exponentially to zero in about 30 minutes. If the line is extended beyond 30 minutes, the time necessary to achieve various levels of assurance can be determined. For example, a level of assurance of 10^{-6} would require about 50 minutes. The D-value for this population at 160°C is about 3.5 minutes; that is, every 3.5 minutes, the population is reduced by another log, or 90 percent.

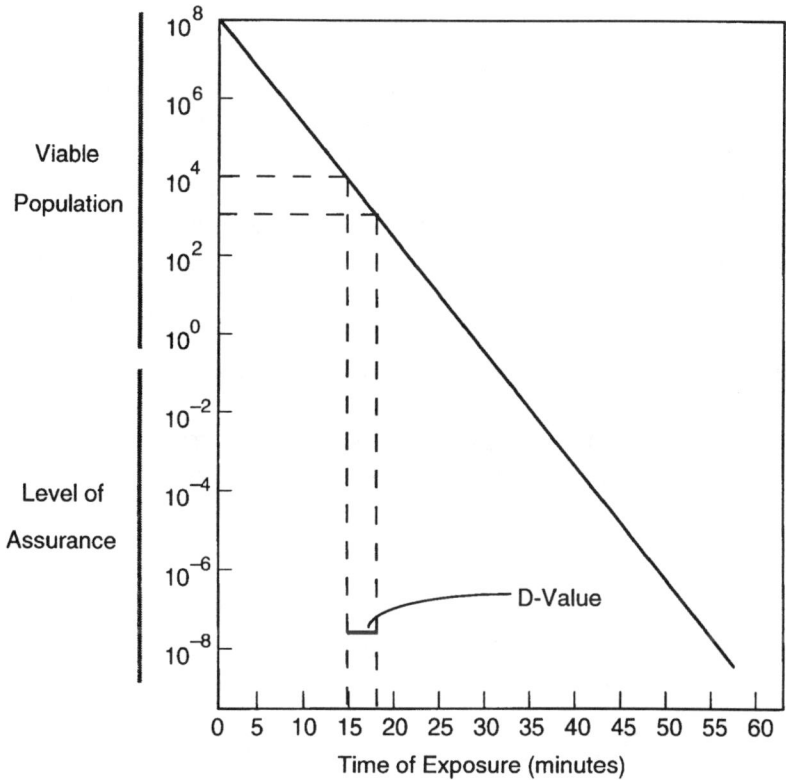

like Figure 3.1 point up an important lesson: when microorganisms are exposed to any lethal agent, whether it is heat, radiation, or chemicals, they do not die instantaneously. It takes a certain amount of time for the agent to kill the entire population.

For a given killing curve, we can determine the DECIMAL REDUCTION TIME, or D-VALUE, which is the time, at the stated temperature, that is necessary to reduce the microbial population by 90 percent, or one log, such as from 10,000 to 1,000. In our example in Figure 3.1, the D-value is about $3^1/_2$ minutes.

By extending the killing curve of the surviving population to less than one cell, we can determine the time necessary to sterilize the specific population in question at that temperature. The distance below one cell the sterilization is carried is considered the LEVEL OF ASSURANCE of sterilization, or OVERKILL. The usefulness of the D-value is in estimating how much time would be necessary to achieve a particular level of assurance. For instance, with a bioburden of 10^8 bacteria and an overkill of 10^{-6}, fourteen D-values would be necessary, which in our example in Figure 3.1, comes out to be about 50 minutes.

Because it is usually not possible to prove that a load of objects is sterile without compromising its sterility, an alternate approach to affirm sterility is to designate a level of assurance of sterility for the load. For example, if 10,000 bottles of saline are sterilized, rather than demonstrate directly that each bottle is sterile, it might be shown that the bottles were subjected to a sterilization process that resulted in a level of assurance of, for example, 10^{-4} ($^1/_{10000}$). That means we could expect no more

---

TABLE 3.1.
SOME FACTORS THAT INFLUENCE
THE GERMICIDAL EFFECTS OF HEAT

Age of cells

Form of cells (vegetative or spores)

Presence of organic matter (blood, grease)

Presence of moisture

Nature of atmosphere (oxidizing or reducing)

pH

than one bottle in 10,000 to contain a surviving organism. To carry out such a calculation, we must of course have a good estimate of the bioburden of the bottles.

Methods of determining bioburdens are discussed in chapter 5. Clearly objects with low bioburdens will require less time to sterilize than heavily contaminated ones. This rule has many practical considerations: unnecessarily long times in the sterilizer reduce throughput and raise costs both in terms of tying up products and process equipment and increasing energy consumption. Furthermore, many products cannot tolerate excessive exposure to heat or sterilization agents. Consequently, raw materials for the production of items that are to be sterilized must be assayed for microbial contamination before their use and be protected from further contamination. If the materials are found to be excessively contaminated, appropriate action must be taken.

Under some circumstances a straight-line killing curve may not be achieved, but rather it may appear to be composed of two or more segments (Figure 3.2). This type of curve is usually caused by the presence of a mixed population, that is, a mixture of several species having varying susceptibilities. Mixed populations are frequently encountered in naturally contaminated materials such as raw materials for bacteriological media, solutions, or

Figure 3.2. Two-Phase Killing Curve. A representative killing curve involving a mixture of two populations, one more resistant than the other resulting in a dog-legged curve.

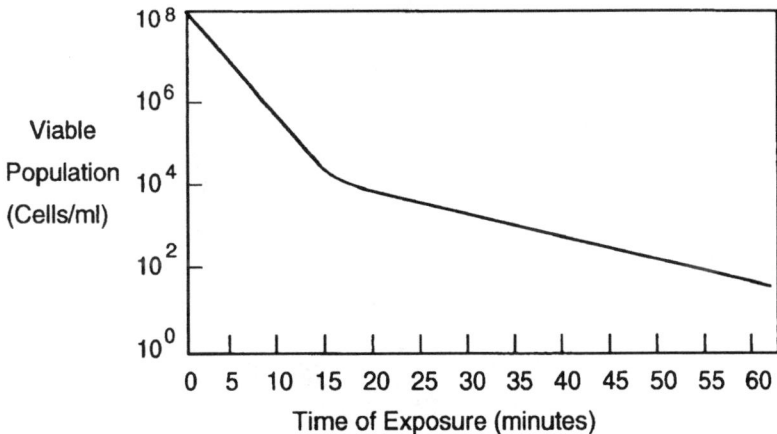

medical products. Therefore, standard sterilization procedures normally must have built-in overkill factors that accommodate worst case situations, that is, the possible presence of moderate numbers of highly resistant bacterial species and spores.

Another useful type of killing curve is the THERMAL DEATH TIME curve (Figure 3.3). Here the same population of cells, say 10^7, is subjected to various temperatures and the minimum time necessary to sterilize the population is determined for each temperature. The thermal death time is defined as the time necessary to sterilize a given population of microorganisms at a given

Figure 3.3. Thermal Death Time Curves. Killing curves of the same organism are prepared for several temperatures. From the curves a D-value for each temperature is determined. For example, at 180°C the D-value is about one minute; at 120°C it is four minutes.

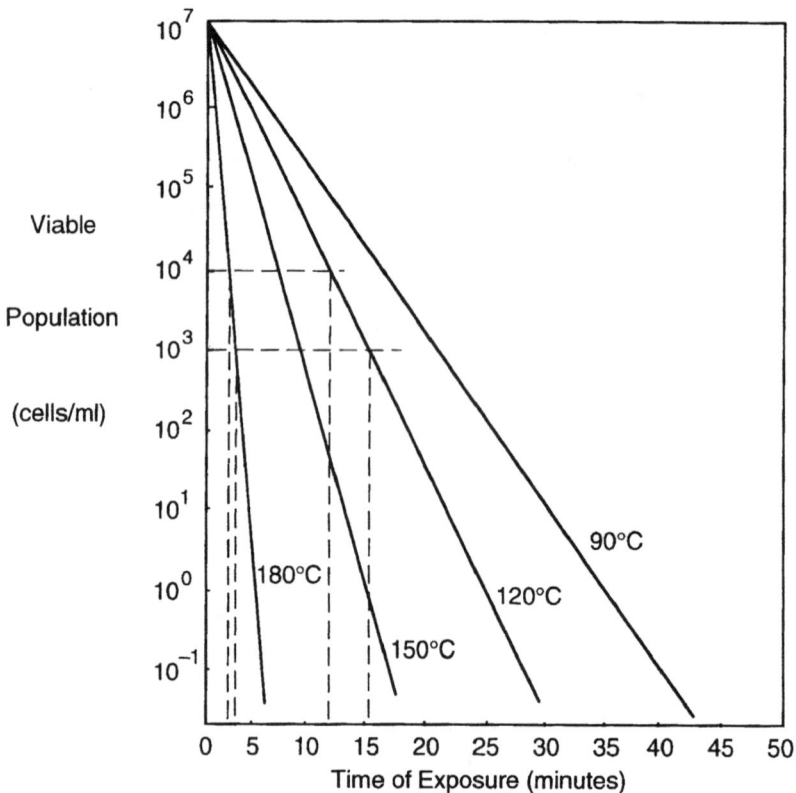

temperature. The D-values derived from the thermal death time curves are useful when determining the cumulative effects of the prolonged heating and cooling down of very large loads when heat sterilized. This will be discussed in detail in the section on dry heat sterilization.

Moist heat sterilization

Moist heat is defined as heat that is accompanied by 100 percent humidity. For sterilization of small laboratory items, the use of moist heat is the most popular method and is normally applied in the form of steam under pressure in an AUTOCLAVE (Figure 3.4). Two types of steam autoclave are used for sterilization: gravity displacement (the most common) and prevacuum. In the gravity displacement autoclave, the lighter steam displaces or forces out the heavier air from the autoclave chamber through a vent at the bottom of the chamber. One hundred percent humidity cannot be achieved if appreciable amounts of air remain in the chamber. With the prevacuum autoclaves most of the air in the chamber is removed by a vacuum pump before the steam is introduced.

We must remember that it is not the heat in the autoclave alone that kills microorganisms, but the combination of heat and saturated moisture. Consequently, items to be sterilized in the autoclave must be ones that allow the steam access to all internal spaces and surfaces. Glassware must have loose-fitting caps or cotton plugs, and wrapped items must be in gauze, Kraft paper, or other material that is permeable to moisture but not penetrable by microorganisms when the items are subsequently removed and handled.

Dense items, such as towels, uniforms, and bandages must be loosely stacked in the autoclave to allow complete steam penetration. Items that have water sealed in them, such as ampoules or serum bottles, may be adequately sterilized in the autoclave, but certain areas inaccessible to the steam, such as between the layers of double seals of caps, may escape sterilization and remain contaminated. Large, empty vessels such as flasks and bottles, even though they may be loosely capped, may also escape sterilization internally if steam is prevented from fully entering due to the presence of (heavier) air in the vessel. Such containers should be placed on their sides to allow the steam to displace the air. This situation is less of a problem with the prevacuum autoclave. Contaminated waste sealed in plastic biohazard bags will not be sterilized unless there is water present in them. Bags that do not contain appreciable amounts of water should either be left open during autoclaving, or have about 100 ml of water added before sealing.

Dense powders and oily substances (sand, talcum powder, paraffin wax, or petroleum jelly) cannot be sterilized in the autoclave because the steam cannot penetrate these materials.

Figure 3.4. A Steam Autoclave. Small objects and volumes of growth medium up to about one liter are sterilized in 20 minutes at 121°C (15 psi).

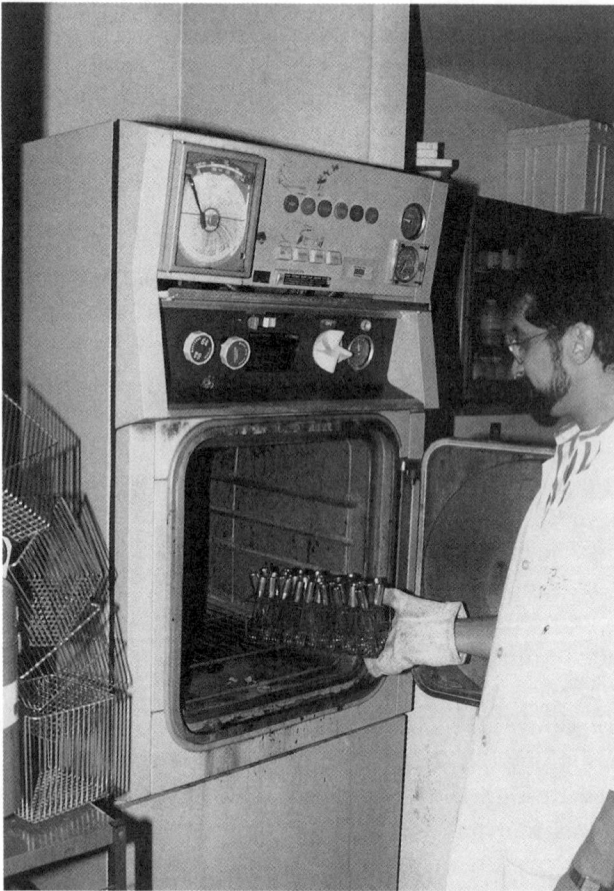

They must be sterilized by dry heat. Also, most types of plastics cannot be autoclaved without significant distortion, but some hold up reasonably well. Always check manufacturers' literature before attempting to autoclave any items that are wholly or partially composed of plastic. Most types of rubber survive autoclaving well but begin to show signs of deterioration following repeated treatments.

Twenty minutes' exposure in the gravity displacement autoclave at 121°C (15 psi) is generally adequate to sterilize a load of small to medium size items, such as surgical instruments and solutions or bacteriological media in volumes of up to about one liter. If sufficient steam pressure is available and the objects can withstand the greater heat, shorter processing times are possible by operating at 136°C (46 psi). At this temperature, only 10 minutes would be required for sterilization.

The autoclave manufacturer's recommendations should be consulted when sterilizing bulky items and large volumes of liquids, since these require additional time to reach sterilization temperature. As a safety margin, additional time (at least two to threefold over normal) should also be applied to heavily contaminated material, such as discarded cultures from environmental samplings or ordinary laboratory cultures. This is because of the high numbers of microorganisms contained in such materials.

All modern autoclaves have a setting marked either "slow exhaust" or "liquids." When using this setting, the steam pressure within the chamber is released gradually at the end of the sterilizing cycle. This prevents liquids from boiling over when the chamber pressure is suddenly reduced after having been heated to well over the liquids' normal boiling point. Nonliquid items such as empty glassware and instruments may be subjected to "fast exhaust," and perhaps to a drying step where a vacuum is applied to the chamber to draw off excess moisture.

PASTEURIZATION is an example of the use of heat to decontaminate rather than sterilize. When milk is pasteurized, it is subjected to 72°C for 15 seconds, or alternatively to 63°C for 30 minutes, a treatment that assures the destruction of all pathogenic organisms commonly known to occur in milk but does not appreciably affect flavor. As a bonus, 90–99 percent of the spoilage organisms in the milk is also destroyed, thereby extending shelf life. Beer, fruit juices, and other fluid food products are also frequently pasteurized to reduce spoilage.

Dry heat sterilization

Dry heat sterilization involves the application of heat at humidities less than 100 percent. The use of dry heat sterilization is limited to items that can withstand the effects of two or more hours at 160°C to 180°C, the temperature range normally used in dry heat sterilization. Clean, empty glassware, some metal objects, powders, and oily materials are most commonly sterilized by dry heat; items containing volatile liquids, plastic, or rubber usually are not. The volumes of oils and powders must be kept to a minimum to allow adequate heat penetration. Layers not over $1/4$ inch in thickness are recommended. Cotton plugs on glassware generally survive without discoloration if not heated above 160°C. Table 3.2 lists some common conditions and limitations for dry heat sterilization. Dry heat sterilization can be carried out in any oven device that can be held at 160°C to 180°C.

In large scale industrial sterilization operations utilizing either moist or dry heat, calculating the time necessary to sterilize large volumes of medium, bulky pieces of equipment, or packaged products can become quite complicated. Figure 3.5 shows a typical situation—the sterilization of a large load of bottles. Note that it takes several hours for the load to reach the sterilization

TABLE 3.2.
EXAMPLES OF RECOMMENDED CONDITIONS FOR DRY HEAT STERILIZATION

Items	Conditions and Limitations
Hypodermic needles	2 hours at 160°C; in tubes with cotton plugs; no stylets (inner wire)
Glass syringes	2 hours at 160°C; plunger and barrel separate, wrapped in muslin
Surgical instruments	1 hour at 160°C; must be clean and on a metal tray
Petroleum jelly	2 hours at 160°C or 1 hour at 170°C; in not more than $1/4$ inch layer
Talcum powder	2 hours at 160°C or 1 hour at 170°C; in not more than $1/4$ inch layer

temperature, and then additional hours for it to cool to room temperature. During those times some killing is occurring, which can be factored in when determining the total time required to sterilize the load. This determination can shorten the overall time the bottles actually need to be exposed to the maximum temperature, saving both time and energy.

This determination is done by assuming the temperature rises in steps. For example, suppose we visualize steps that are 10 minutes wide, and then determine the average temperature of each

Figure 3.5. Sterilizing a Large Load. Time-temperature profile of a large load of product undergoing heat sterilization. While the oven heats up quickly (narrow line), the load (heavy line) takes about seven and a half hours to reach the maximum temperature (measured in the center of the load), and three hours to cool. However, some microbial killing is occurring both during the heat-up period (A) and the cool-down period (C), the amount of which in D-values can be determined with the aid of the thermal death time curves (Figure 3.3). These D-values, added to the D-values during the plateau (B), will yield the total amount of killing the product was exposed to. For example, suppose during period A the load accumulated three D-values, six during period B, and one during period C, totaling ten D-values, which would be sufficient to sterilize a load with a bioburden of 10^4 to a level of assurance of 10^{-6}.

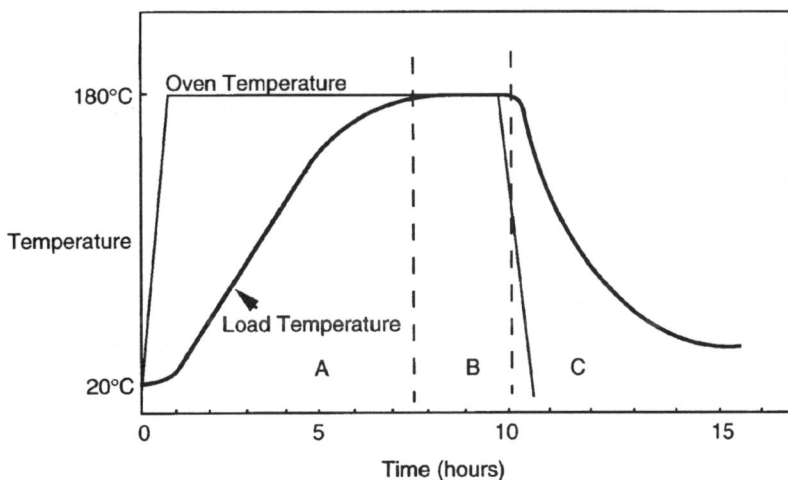

step. From thermal death curves prepared previously (Figure 3.3) we can estimate how much killing, in terms of D-values, occurs at each step, which is then added up for both the heat-up and the cool-down portions of the curve. What remains of the bioburden (plus that required for overkill) is then taken care of at the maximum operating temperature.

Radiation

Ionizing radiation

Certain types of radiation have been useful in the control of microorganisms. IONIZING RADIATION is radiation of sufficient energy to cause the ionization of atoms. Alpha, beta, gamma, X-radiation, and electron beams are all examples of types of ionizing radiation. Because of the relative ease of producing them, gamma radiation and electron beams are the most common of the ionizing types of radiation for the control of microorganisms, the principal application being the sterilization of drugs, disposable plastic medical devices, and bacteriological laboratory ware. Though controversial, ionizing radiation is increasingly being used to decontaminate food products.

Typical doses required to sterilize objects are generally of the order of 2.5 million rads, which can be delivered in a few seconds by present day radiation facilities. A RAD is that amount of radiation that results in 100 ergs of energy absorbed by one gram of matter. (As a comparison, a typical chest X-ray delivers less than 0.1 rad to the body. A lethal dose for a human is estimated to be about 400 rads.) ^{60}Co is frequently the source of the gamma radiation, and electron accelerators, giant versions of the electron gun in a television picture tube, provide the electron beams.

Radiation sources are installed in heavily shielded rooms to protect the operators from radiation, and products to be sterilized are moved through the radiation beam on a moving belt, on carts, or hanging from a monorail. Because of its great power of penetration, ionizing radiation can be used to sterilize items packed in bulk or case quantities. Contrary to popular opinion, objects sterilized in this manner do not themselves become radioactive.

Nonionizing radiation

Electromagnetic energy of a wavelength in the range of 240 to 280 nanometers (nm) with a peak at 260 nm is highly lethal to microorganisms. This form of radiation is in the ULTRAVIOLET region of the light spectrum and is of the nonionizing type. The most convenient source of this type of radiation is low pressure mercury vapor, or germicidal lamps commonly seen in microbiological laboratories. These lamps emit a large proportion of their energy at 254 nm. Germicidal lamps should not be confused with so-called black lights or Woods lamps, which emit ultraviolet light principally in the 320 to 420 nm range. These lamps are intended to activate fluorescence in mineral specimens, chromatography preparations, and other applications and do not produce significant radiation below 300 nm. The different types of UV lamps can be distinguished by the appearance of the lamp envelope: black lamps have a dark envelope while the germicidal lamps have a clear envelope.

Except under carefully controlled conditions, ultraviolet light is not a reliable sterilizing agent and generally can only be used for decontamination. This is because of two characteristics of UV light: it follows the inverse square law, and it is not very penetrating. The inverse square law refers to the fact that as a target moves away from the source of radiation, the intensity of the radiation falls off in a manner that is inversely proportional to the square of the distance. That is to say, if a bacterium moves from 1 inch to 2 inches from a UV lamp, the intensity of radiation that cell feels has been reduced to $(1/_2)^2$ or $1/_4$ of what it felt at 1 inch. What that means on a practical basis is that the target organisms must be relatively close to the UV lamp to receive significant radiation.

A second characteristic of UV light is that it does not penetrate dirt, dust, glass, most plastics, most bacteriological media, or more than a few micrometers of cytoplasm. Thus, UV light generally cannot be used to sterilize dirty or heavily contaminated objects, solutions in glass or plastic containers, or solutions that are cloudy or contain organic material or dense microbial populations. UV light is successfully used to sterilize clear water or the surfaces of clean objects. It can be used to sterilize the interiors of clean, empty bottles of low density polyethylene provided the radiation source is of sufficient intensity. Despite these shortcomings, UV light has some useful applications in the laboratory. One of its most common uses is in the decontamination of surfaces and confined air spaces such as the interiors of biohazard and transfer hoods and air ducts. UV is also used to control microbial activity in water systems.

Great care must be practiced in the presence of operating germicidal lamps as their radiation is highly damaging to the eyes and skin. In a few seconds a high intensity UV lamp is capable of delivering a painful sunburn that would take several hours to obtain at the beach. Gloves, long sleeves, and a full face shield should be worn whenever work near operating UV lamps is necessary.

The output of an ultraviolet lamp should be monitored on a regular basis, because such lamps have a tendency to deteriorate without warning. In addition to normal aging, the accumulation of dirt and grease on the outer surface of the lamp can severely reduce delivery of the lethal wavelengths. UV light meters are available for monitoring germicidal lamps (Figure 3.6).

Microorganisms subjected to radiation generally follow exponential kinetics similar to what has been described for killing by

Figure 3.6. An Ultraviolet Intensity Meter.

heat, allowing one to predict the effects of a particular dose of irradiation on a given population of microorganisms.

Low Temperatures

Storage of perishable items, such as foods or biochemical solutions, at low temperatures is an excellent and widely used method of inhibiting microbial activity. If the storage temperature is between 0°C and about 5°C or so, the technique is referred to as refrigeration; below 0°C is freezing. A typical refrigerator maintains a temperature of about 4°C, at which point mesophilic microorganisms cannot grow. Psychrophiles, on the other hand, grow slowly, even at 0°C, which is why refrigerated foods do eventually spoil. A household freezer may go as low as –20°C, while freezers made for laboratory use can reach –80°C. Growth is not possible at these low temperatures. Although some viability may be reduced, exposure to low temperatures is not a reliable method of decontamination. Quite the contrary, microbial suspensions are frequently preserved for later use by freezing.

Filtration

Liquids and gases can be reliably sterilized by passing them through various types of filters made of glass, ceramic, or organic polymer. The organic polymer filters are the most popular type in microbiology laboratories and manufacturing applications. They commonly consist of membranes of cellulose nitrate or acetate, or Teflon (Figure 3.7). For sterilization, the porosities of the filters must be such that they retain the smallest bacterial cells but still allow reasonable flow-through. An effective porosity of not more than 0.22 μm is considered the most reliable for removing the smallest bacteria. Most viruses are not trapped by these filters. If viruses must be eliminated, filters of smaller porosities or other methods of sterilization must be used.

A particular type of filter for air that is familiar to cleanroom workers is the High Efficiency Particulate Air or HEPA filter (Figure 3.8). These filters, usually made from a fiberglass paper, remove from 99.95 percent to 99.995 percent of airborne particles 0.3 μm and larger. Although HEPA filters do not sterilize the air, their use in laboratory and cleanroom air supplies and transfer and biohazard hoods drastically reduces airborne particulate contamination and provides considerable protection for products and for laboratory personnel. These types of filters are discussed further in chapter 4.

Chemical Methods

Sterilants

The ideal chemical sterilizing agent should have at least the following characteristics:

- Sufficient toxicity toward microorganisms, including bacterial spores, to kill large numbers in a relatively short period of time

- Little or no toxicity toward humans

- Good penetrating power

- Little or no residue

The ideal chemical sterilant does not exist, and only a few chemicals come sufficiently close to fitting these criteria to be

Figure 3.7. Sterilization by Filtration. A solution can be sterilized by passing it through a membrane filter with a porosity of 0.22 μm.

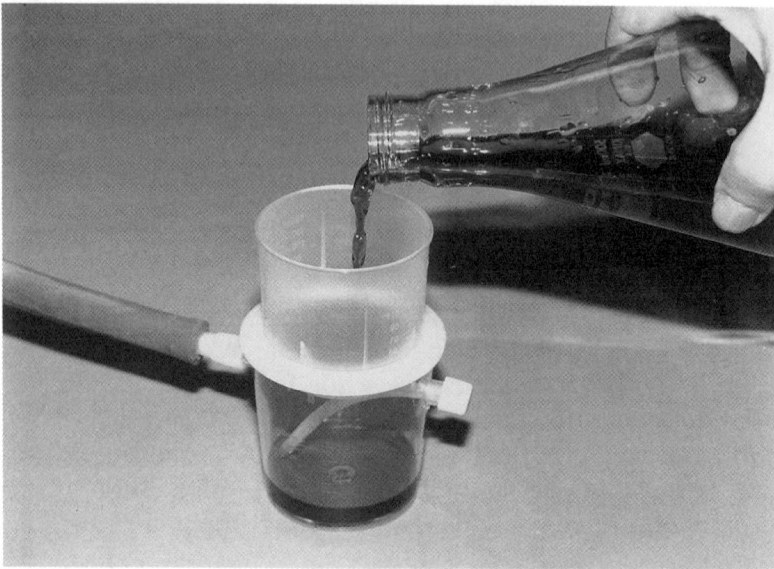

Figure 3.8. A HEPA Filter Unit. A complete unit (top) is ready for installation. It measures 2 ft by 2 ft. A section of a HEPA filter (bottom) showing the corrugated aluminum spacers and the zigzag arrangement of the filter medium. Arrows indicate the direction of air movement through the filter.

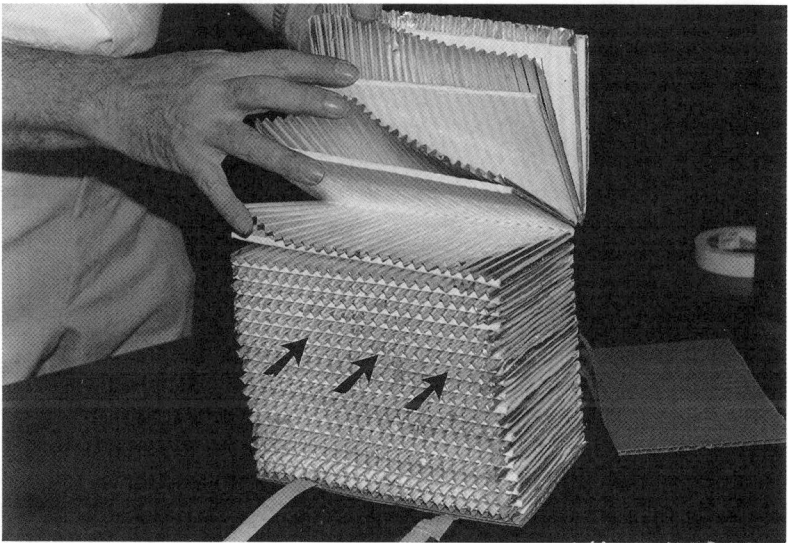

considered reliable enough for wide use. Some of these are formaldehyde, ethylene oxide, glutaraldehyde, and chlorine dioxide.

FORMALDEHYDE is a colorless, highly toxic gas with a sharp, stinging odor. It is very soluble in water. A 37–40 percent (by weight) solution is known as formalin. Formaldehyde can be used either as a solution or in the gaseous form. The gaseous form is widely used to sterilize air spaces such as laboratories, hoods, and air handling ducts. The gas is most conveniently generated by heating the solid (flake) form of the chemical, paraformaldehyde, on an electric hot plate at 230°C in the space to be treated. The volume of the space is estimated and then completely sealed with heavy plastic sheeting to prevent loss of the gas and danger to personnel in the surrounding facility.

The recommended amount of paraformaldehyde to use to sterilize a specific air space is 300 mg for each cubic foot of space to be treated. Optimum conditions are a temperature of about 20°C to 22°C and a relative humidity of 60–80 percent. The humidity can be raised by generating steam in the space just before the formaldehyde is released. This is accomplished by boiling a quantity of water on a second hot plate. Following treatment, the space must be thoroughly ventilated before it is used. Generators for sterilizing air spaces with formaldehyde are also available. They have the additional feature of releasing a neutralizer following the sterilization thereby allowing an earlier occupation of the exposed space. The future use of formaldehyde remains in doubt as there is some evidence that it is a carcinogen.

ETHYLENE OXIDE (ETO) is another gas, but unlike formaldehyde, it has limited solubility in water. Being highly flammable, it is frequently mixed with carbon dioxide or a fluorocarbon to reduce the fire hazard. It is an excellent general sterilant with an enormous ability to penetrate plastic or paper wrappings. Products such as bacteriological laboratory ware and medical devices may be sterilized by ETO in their final bulk packaging, but afterward they must be subjected to a lengthy quarantine period to allow for complete outgassing. ETO sterilization is carried out in airtight chambers similar to steam autoclaves. Many commercial ETO chambers may exceed 3000 ft^3 in volume. Because of the time necessary for the gas to penetrate the innermost areas of a load of items, ETO sterilization can be slow, requiring 24 hours or more for a large load.

Growing evidence of the carcinogenicity of ETO has curtailed its use where repeated exposure to workers may occur. Use of ETO that is mixed with fluorocarbons has also been under criticism because of the damage fluorocarbons cause to Earth's ozone layer.

ETO also is used to sterilize many nonmedical items such as imported spices, cosmetics, and museum artifacts.

CHLORINE DIOXIDE has recently been added to the short list of reliable chemical sterilants. It is a gas but almost impossible to use as such because of its toxicity and unpleasant odor, which has been described as a mixture of chlorine and fuming nitric acid. Straight water solutions of chlorine dioxide are unstable, but stabilized solutions of it are available for laboratory and industrial use.

GLUTARALDEHYDE is frequently used in diluted (around 2 percent) solutions to sterilize small medical devices, catheters, and the like, but they must be rinsed with sterile water before use. Glutaraldehyde has the capacity to kill bacterial spores but only after extended exposure. Because of its cost, its use is limited.

Disinfectants

Many chemical agents are not sufficiently reliable to act as sterilants but are still useful in the control of microorganisms. These are the disinfectants and sanitizers. Names of classes of chemicals that kill microorganisms frequently end with *-cide,* with the root of the word referring to the kind of organism affected. Thus, agents that kill bacteria, (vegetative forms only), fungi, or viruses are called bactericides, fungicides, or viricides, respectively. Considerable overlapping occurs. For instance, many bactericides exhibit fungicidal and viricidal activity as well. Agents that are sufficiently powerful to kill bacterial spores are called sporicides.

If an agent only prevents growth of the organisms but does not kill them, the suffix *-stat* is used: bacteristat, fungistat, and so forth. There cannot be "viristats," for viruses are not capable of independent growth. Some chemicals are *-cides* at high concentrations but become *-stats* at lower concentrations. Since exposure to *-stats* does not kill microorganisms, growth normally resumes when the agents are removed.

Just as in heat sterilization (Table 3.1), many factors alter the effectiveness of chemical sterilants and disinfectants. Concentration and temperature are parameters that are under the control of the user. As a general rule, elevated temperatures improve the effectiveness of chemical agents provided the agents do not decompose or evaporate. Increasing the concentration of an antimicrobial agent will not necessarily improve its effectiveness. More is better is not always the rule. For example, 70 percent ethanol is more effective as a disinfectant than 100 percent. It is imperative that manufacturers' instructions be followed carefully with regard to the proper concentration for specific applications. Also, some disinfectants are inactivated by the mineral content of hard water and must be diluted with distilled water. Of greater concern is the fact that tap water has been known to contain bacteria that are capable of growing in many types of disinfectant solutions.

Other factors that severely impact the effectiveness of a disinfectant include the type and age of the microorganisms that are to be eliminated. As has been pointed out, bacterial spores can be as much as a million times more resistant to chemical agents than vegetative cells. An object heavily contaminated with bacterial spores will not be significantly disinfected by most chemical agents. Also, objects heavily soiled with organic material, such as blood and fecal matter, require considerably longer exposures. Obviously, objects that are to be decontaminated with any reliability must be thoroughly precleaned.

QUATERNARY AMMONIUM compounds ("quats") are popular disinfecting agents. Examples are benzalkonium chloride and cetyl pyridinium chloride. Being cationic detergents, they owe some of their antimicrobial activity to the physical removal of cells, but they are also moderately effective bactericides in their own right, and at concentrations as low as 0.0005 percent they can still be bacteristatic. Their action appears to involve the disruption of cell membrane permeability. Most popular mouthwashes contain these agents, which attests to their low toxicity for humans. However, quats also show low toxicities toward certain types of bacteria, such as pseudomonads and tuberculosis bacilli; consequently, they cannot be relied on to disinfect medical instruments and similar critical applications. Quats can be inactivated by anionic detergents and surfactants (wetting agents), meaning that products containing quats must never be mixed with or alternated with other products containing these interfering materials. Quats must always be diluted in deionized or distilled water, because they are also inactivated by hard water.

PHENOL (carbolic acid) is an effective disinfectant, but its toxicity and corrosiveness have severely limited its general use. Derivatives of phenol, known as phenolics, such as o-phenylphenol, are considerably less toxic, actually more effective than phenol, and have become very popular for decontaminating equipment, furniture, and floors. Hexachlorophene, another derivative of phenol, is a good skin antiseptic, but its use must be carefully controlled. It is absorbed through the skin and is highly damaging to the central nervous system of newborn children. The antimicrobial action of the phenolics appears to be related to their ability to precipitate proteins, but even at very low concentrations when precipitation would normally not occur, the phenolics retain some antimicrobial activity. This appears to be related to their additional capacity to disrupt cell membrane function, preventing essential nutrients from entering the cell or causing leakage of cell contents. The phenolics are normally not compatible with nonionic detergents.

ALCOHOLS, such as ethanol and isopropanol, are effective and widely used as disinfectants and antiseptics. As in the case of the quats, the action of the alcohols appears to be a combination of cleansing action and chemical activity. Being organic solvents, the alcohols extract lipids from cell membranes and thereby destroy the membranes' functioning. Water solutions of about 70–90 percent (by volume) of ethanol or isopropanol are bactericidal against vegetative cells but have little effect on bacterial spores. Another limiting factor of alcohols is their volatility. Because they have a tendency to evaporate quickly, exposure time is often cut short unless they are applied generously.

CHLORINE and IODINE belong to a group of chemical elements known as halogens. Chlorine is a gas at room temperature and is commonly used as a disinfectant of water for domestic use, although its use is being drastically reduced because of evidence that it may be responsible for the formation of carcinogens. Organic chlorine compounds known as chloramines have replaced gaseous chlorine in many applications. Inorganic chlorine compounds, principally hypochlorites (ordinary laundry bleach), are also widely used as disinfectants. In the presence of water, hypochlorite (OCl^-) slowly releases nascent oxygen, a powerful oxidizer that is highly microbicidal. Its principal effect seems to be the oxidation of sulfhydryl groups that are essential for the functioning of many proteins.

Common applications of chlorine and its compounds as disinfectants other than for water supplies are as sanitizers in the food handling industry, including dairies and restaurants. Chlorine compounds (hypochlorite) tend to be unstable, resulting in short shelf lives. Working solutions should always be prepared fresh daily. Dilutions of bleach of 1:100 are useful disinfectants on clean surfaces, but if significant amounts of soil are present, a 1:10 dilution is recommended. Hypochlorite breaks down in the presence of organic matter so that heavily soiled objects must be cleaned before decontamination. It has some sporicidal activity, but an insufficient amount to be considered a reliable sterilant.

Iodine has been a familiar household antiseptic for generations, usually as an alcoholic solution known as a tincture. Iodophors, organic compounds of iodine, have widely replaced tincture of iodine for the treatment of wounds, for they are less irritating to open tissue and have thus gained the nickname "ouchless" iodine. Iodine and its compounds are good skin antiseptics, being bactericidal and fungicidal, and are widely used on patients before surgical procedures. Other uses include disinfection of water and food handling equipment.

Other than the applications cited above, widespread industrial use of the halogens for the control of microorganisms has been limited because of their corrosiveness toward some metals.

HYDROGEN PEROXIDE has long been a popular antiseptic and disinfectant. Like chlorine and its compounds, it is a strong oxidizer. Used in concentrations of 3–6 percent, it is a reliable disinfectant *on clean objects*. It is unstable in the presence of organic debris. It is also used as a vapor in certain applications.

Selection of disinfectants

There is no such thing as the ideal disinfectant. The commonly used disinfectants are active against vegetative microbial cells, but few are sporicidal. Truly sporicidal agents are usually too toxic for general use and are limited to special applications. Certain disinfectants are more appropriately used on floors and walls, while others better lend themselves to the decontamination of process equipment. Cost is frequently a factor. Some very effective agents, such as glutaraldehyde, are too expensive to use in large volumes for floors and walls, but are ideal for small pieces of equipment or surgical instruments.

About 10,000 disinfectants are registered with the Food and Drug Administration, which closely monitors their labeling. However, what does not appear on a label is frequently more important than what does appear. Workers should not assume capabilities of a disinfectant unless they are specifically and unequivocally expressed on the label or in the manufacturer's literature. A product that is labeled simply as a "germicide" or "disinfectant" without any further clarification may only be effective against relatively sensitive bacterial species but useless against tuberculosis bacteria and *Pseudomonas.*

It must always be assumed that a given disinfectant is not sporicidal unless the manufacturer's literature specifically states that it is. Sporicides must be clearly labeled as such. Beware of products that claim "kills spore-forming bacteria" or words to that effect. This simply means that they kill the vegetative forms of the spore formers. Nor can we rely on products to kill viruses, fungi, or tubercle bacilli unless the label specifically states that they can. In all cases, look for results of standard tests that back up the claims and then validate the products yourself using the same or similarly recognized tests.

Validation of disinfectants

The two most commonly cited validation tests for disinfectants are the phenol coefficient and the use dilution test. The PHENOL COEFFICIENT is basically a comparison of the bactericidal ability of a disinfectant with that of phenol. Thus, a phenol coefficient of 1.0 denotes that the disinfectant is essentially equal to phenol; coefficients greater than 1.0 mean it is equal to phenol at proportionally lower concentrations. Put another way, if a product is reported to have a phenol coefficient of 10, it means that a 0.1 percent solution of the disinfectant is as good as a 1.0 percent solution of phenol.

The phenol coefficient has been largely replaced by the USE DILUTION TEST. In the phenol coefficient test, various dilutions of a disinfectant are compared with the same dilutions of phenol, but the concentrations of the product used in the test frequently have no relationship to the concentrations actually used in practice. In the use dilution test, actual in-use concentrations of a product are tested for their bactericidal activity. Under the conditions of the test, if the challenge bacteria are killed in 10 minutes' exposure to the in-use concentration, the disinfectant passes the test.

Rotation of disinfectants

It is generally assumed that if we were to use the same disinfectant for extended periods of time, the resident population of microorganisms in a facility would eventually develop resistance toward the agent. By alternating disinfectants, it may be possible to avoid this problem. However, unless one has strong evidence that such resistant strains are appearing, switching to another disinfectant may cause more problems than staying with the same one. One reason for this is that many types of disinfectant are incompatible. Nearly all disinfectants leave a residue on surfaces on which they are used. If one were to switch to an incompatible agent on those surfaces, yesterday's residue might neutralize today's disinfectant or, at best, the two agents may form an insoluble film that would make cleaning difficult. It is therefore prudent to test resident microbial populations regularly for signs of resistance and, only if present, select a compatible, substitute disinfectant.

Validation of Sterilization

How does a company know if its products are being adequately sterilized? It is obviously not possible to test every unit, for the very act of testing would compromise its sterility. A certain number of units may be selected on a statistical basis to be tested on the assumption that they represent the entire lot of units.

As a further assurance, the sterilization PROCESS must be monitored to determine that all equipment and procedures associated with sterilization are operating properly. Various monitoring methods can be used, such as placing electronic heat sensors called thermocouples within a load of items to be sterilized to confirm that the items have reached the proper temperature. This is generally acceptable with dry heat sterilizers, but, as we have pointed out, moist heat sterilization depends on the combination of heat and moisture. Consequently, an item may have reached 121°C in an autoclave, but unless the correct amount of moisture is present, the item may not be sterilized.

In monitoring gaseous chemical sterilization, remote sensors can report concentration of the agent, humidity, or temperature as indicators that sterilization conditions were met. In addition, one may use paper strips that are impregnated with chemicals that change color when exposed to ethylene oxide, for example, confirming that the sterilant had reached the indicator. These are known as CHEMICAL INDICATORS.

BIOLOGICAL INDICATORS offer an additional degree of confidence that a given sterilization step is working. As the term suggests, biological indicators involve the use of viable organisms to challenge the sterilization operation. Standardized, commercially prepared indicators are available that consist of known numbers of bacterial spores that are either placed on strips of filter paper or in vials. The spores originate from species that are known to be exceptionally resistant to the specific type of sterilization to be tested: moist or dry heat, gaseous, or radiation sterilization (Table 3.3).

The indicators are placed in the innermost region of a load of items properly packaged for sterilization, then retrieved following the sterilization procedure and tested for viability. Because the indicators represent a worst case situation in terms of numbers, levels of resistance, and location in the sterilizer, if the indicators have been sterilized, the entire load can be considered sterilized. An exception to this might be if it is suspected that the objects or material to be sterilized are contaminated with organisms that are more resistant to the sterilization conditions than those in the indicator. If this is the case, then the contaminating organism must be isolated and used as the biological indicator for sterilizing the objects. In the same vein, if the numbers of contaminating organisms on the objects to be sterilized exceed those in the

TABLE 3.3.
SOME COMMON BIOLOGICAL
STERILIZATION INDICATORS

Sterilization Method	Biological Indicator (Spores)
Moist Heat	*Bacillus stearothermophilus*
Dry Heat	*Bacillus subtilis* var. *niger*
Ethylene Oxide	*Bacillus subtilis* var. *niger*
Gamma Radiation	*Bacillus pumulus*

biological indicators, appropriate adjustments of sterilizing conditions must be made as well.

A recent incident points up the need for constant and comprehensive monitoring and testing of sterilization operations. Batches of imported cotton contaminated with fungal spores that are particularly resistant to ethylene oxide had been showing up in the U.S. Manufacturers of medical products containing the cotton that were labeled as sterile had to recall large numbers of their products because they contained viable spores. Fortunately, the fungus is harmless, but the incident underscores the need for vigilance and a willingness to expect the unexpected.

Complete documentation that all processes are working properly is known as VALIDATION. Maintaining validation records is a critical part of the plant microbiologist's responsibilities.

Maintaining Sterility

An important element of all sterilization procedures is the maintenance of the sterility of items from the time they are sterilized until they are used. For laboratory glassware to be used immediately, loose-fitting metal foil caps or cotton plugs are adequate, but for items that are to be stored for periods of many weeks or months, additional coverings of foil or paper should be used prior to sterilization to protect the items from accumulated dust. (*Note:* Any coverings must be compatible with the type of sterilization used. Paraffin film would obviously not be appropriate for heat sterilization, and foil must not be placed too firmly on empty vessels for steam sterilization that prevents steam from entering them.) For wrapped items, the reliability of the packaging must match or exceed that of the sterilization procedure. Material for wrapping must not be permeable to microorganisms, and it must be protected from damage. Sterile packages must be carefully examined before the items are used, and any that show the slightest damage must be repackaged and resterilized or discarded. Care must be practiced when the packages are opened so that the objects inside are not contaminated. For certain types of packaging with limited shelf lives, expiration dates must be determined and shown on the items.

Control of Microorganisms in High Purity Water Systems

Many industrial processes require large volumes of extremely pure water. Natural water is generally heavily laden with dissolved salts, gases, organic matter, inert particulates, and microorganisms, all of which must be removed. Typical water purification systems use one or more of the following steps: filtration, distillation, ion-exchange, and reverse osmosis. Industries involved in the manufacture of foods, pharmaceuticals, and semiconductors require water of varying purities. Generally, water that is designated as potable is sufficiently pure for most applications in the food industry. Potable is defined as fit for human consumption. On the other hand, water for injection into the body must be sterile as well as free of particulates and bacterial endotoxins (pyrogens), which are removed by distillation.

The purest water is required by the semiconductor industry. Very Large Scale Integrated (VSLI) circuits, where thousands of electronic components are concentrated in less than a square centimeter, are easily shorted out by a one micrometer particle, like a bacterial cell. During the manufacture of integrated circuits, the wafers (the substrates on which the circuits are assembled) are subjected to a number of chemical treatments, each of which is followed by a water rinse. Unless the rinse water is extremely pure, it can contaminate the circuit and render it inoperative. In the purification of water for use in integrated circuit manufacturing, dissolved solids are removed by reverse osmosis (RO) and ion exchange. Organic contaminants are adsorbed by activated carbon filters, and ultrafilters capture particulates that may have originated from the ion exchange and carbon beds. At specified steps in the purification process, bactericides and ultraviolet light must be used to control microbial activity, for there are some types of bacteria that can grow in purified water. They are frequently found adhering in large numbers to the membranes of the RO system, to the resin beads of the ion exchange units, and to the activated carbon particles.

Hypochlorite can be used to decontaminate the RO membranes, but resin beds are treated with formaldehyde, for chlorine attacks the resin. Peracetic acid has also been used to decontaminate RO membranes. Frequent regeneration of resin beds is also an effective means of preventing microbial buildup. Storage tanks can be made microbefree by maintaining residual chlorine levels around 0.2–0.4 parts per million, but the chlorine must be prevented from getting to the deionizing beds, usually by activated carbon filtration. All vents on storage tanks should be equipped with filters to prevent contamination from entering. Filter media should be hydrophobic to prevent plugging by moisture and microbial grow-through. In addition, a blanket of nitrogen gas held at a slight positive pressure in the storage tanks will inhibit most microbial growth and prevent contamination from entering the tank through leaks in vents, access doors, and other openings. Finally, some systems utilize heating loops through which water is continually circulated to reduce viable microbial populations.

Summary

Many physical and chemical agents are available for the control of microorganisms. Some, such as heat or ethylene oxide, are sufficiently effective to be relied upon to sterilize objects whereas most others are less effective and can only decontaminate. The choice of an appropriate method to use is not always obvious. Consideration must be paid to the compatibility of the method with the materials that make up the items to be treated. Many materials cannot tolerate the heat necessary to sterilize while others are damaged by chemical disinfectants. Some situations do not require sterilization, and decontamination may be sufficient. Raw materials, including water, must be handled carefully to avoid excessive bioburdens prior to sterilization and may have to be precleaned to remove foreign matter that would reduce the effectiveness of the sterilant.

4 Facilities and Personnel Controls

Clean rooms are specifically constructed work areas of a research or manufacturing facility that are designed to assure constant environmental conditions, such as temperature and humidity, and to reduce airborne particulate contamination. Such facilities are generally classified according to the number of airborne particles per unit volume of air of a specific size range that is detected in them. Two classification systems are in use: one is based on English units (particles per cubic foot), which is used only in the U.S., and one is based on the metric (SI) system (particles per cubic meter). Thus, a metric Class M4 clean room generally means that under normal operating conditions, the air in that room contains no more than 10,000 (that is, 10^4) particles per cubic meter 0.5 micrometers and larger. Under the English classification system, a Class 10,000 room means the room has no more than 10,000 particles per cubic foot 0.5 μm in size. Table 4.1 lists the common cleanroom classes in both systems. The latest version of Federal Standard 209 should be consulted for complete definitions of cleanroom classes and additional information.

Evidently, a Class M3 room is cleaner than a Class M4, and a Class M2 is cleaner than a Class M3. The cost of constructing a clean room roughly increases geometrically as one moves up through the cleaner classes; therefore, it is imperative that when a clean room is being planned, specific requirements are well known so that the facility is not overdesigned.

Facilities where concern is also focused on the control of viable microbiological contamination are sometimes referred to as

BIOCLEAN facilities. In discussing the control of particulate contamination in bioclean facilities, we must distinguish between viable and nonviable particles. A viable particle is one that has one or more viable microorganisms attached to it; if such a particle were to be placed in an appropriate microbiological growth medium, microbial growth would be observed. A nonviable particle has no detectable viable organisms on it. Notice the word *detectable* in the definition of nonviable particle. In essence that implies a particle may be nonviable (microbiologically undetectable) on one medium but viable (microbiologically

TABLE 4.1.
COMPARISON OF U.S. AND SI (INTERNATIONAL) CLEANROOM STANDARDS

Class Name		Upper Limits for 0.5 μm Particles	
SI	**FS209E**	**particles/m³**	**particles/ft³**
M1		10.0	0.283
M1.5	1	35.3	1
M2		100	2.83
M2.5	10	353	10
M3		1,000	28.3
M3.5	100	3,530	100
M4		10,000	283
M4.5	1,000	35,300	1,000
M5		100,000	2,830
M5.5	10,000	353,000	10,000
M6		1,000,000	28,300
M6.5	100,000	3,530,000	100,000
M7		10,000,000	283,000

From General Services Administration. 1992. *Airborne Particulate Cleanliness Classes in Cleanrooms and Clean Zones.* Federal Standard 209E. Washington, DC: General Services Administration.

detectable) on another. That is why it is important to use standard media when determining environmental microbial populations. Most cleanroom classifications, such as those in Federal Standard 209, are based on total particle count without consideration of a microbiological load.

Under standard assay conditions, whether a particle is viable or nonviable is generally a matter of how it was created. Smoke particles from soldering or metal grinding operations are usually sterile (nonviable) whereas dust particles from outdoor air, droplets emitted from a cough, or sloughed skin particles are likely to be viable. Automatic particle counters cannot distinguish viable from nonviable particles, and it should be noted that there is little or no correlation between total airborne particle counts, as measured by an automatic particle counter, and the viable particulate level as determined by bioassay. There can be facilities where total particle counts are very high, but viable counts are low, and in other locations viable counts may be very close to the total count. To repeat, it depends on the source of the particles.

Clean Facilities

Certain general features are common in all well-designed clean facilities. In manufacturing areas, the flow of raw materials to final product follows a path of increasing cleanliness. Cleanroom design normally dictates that clean air moves through the room in a unidirectional manner; that is, the supply air ducts are concentrated at one end of the room and exhaust ducts are placed at the other end. Supply and exhaust gratings often make up entire wall areas, resulting in air movement within the room that approaches LAMINAR FLOW. True laminar flow where all air molecules are moving in parallel paths at constant velocity is not possible unless the room is relatively small and completely empty of furniture and other objects. For that reason, the term UNIDIRECTIONAL FLOW is probably more accurate and is therefore preferred when describing air movement in a clean room.

Depending on specific needs, a clean room may be designed for horizontal, vertical, or radial air movement (Figure 4.1). In radial air flow, HEPA-filtered air enters a room either at one junction of a wall and ceiling or the center of the ceiling and exits at

Figure 4.1. Types of Clean Rooms: (top) Vertical flow, (middle) horizontal flow, (bottom) radial or vector flow. Blowers (not shown) recirculate room air through plenum spaces. Unidirectional air movement within clean rooms is shown idealized. A certain proportion of the air is discharged and made up with fresh air.

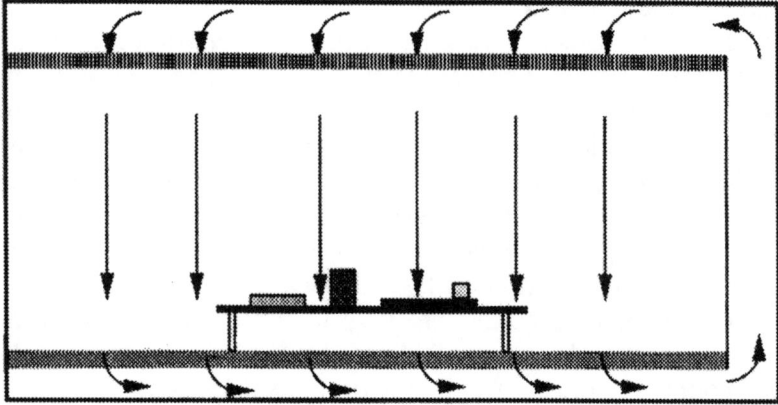

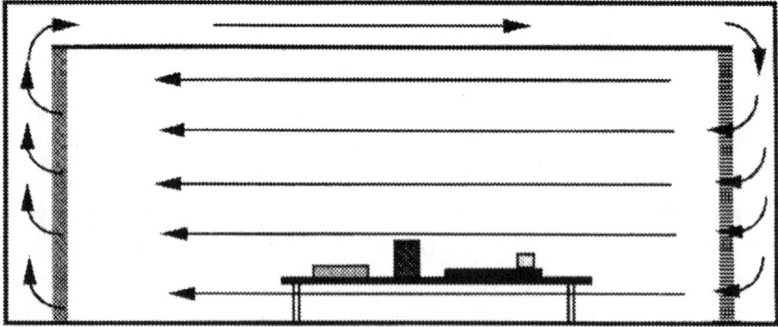

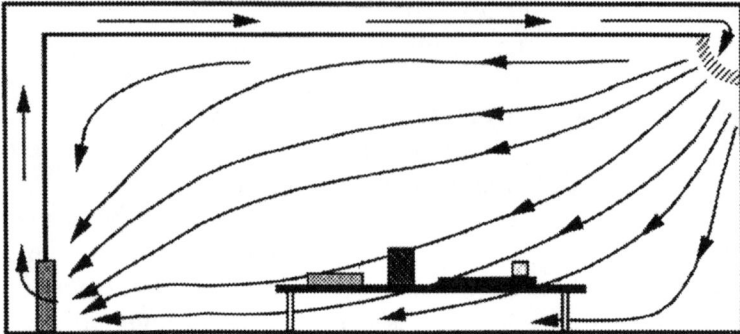

the opposite wall-floor junction. Higher levels of cleanliness may be achieved inside a clean room by installing clean benches or additional sources of clean air which, having their own filters, concentrate on a specific critical area within the clean room. Frequently, these areas can be separated from the main room by heavy plastic strips that contain the cleaner air but still allow movement of personnel.

The basic principle of unidirectional air movement to achieve a clean environment is to supply the room with air that is as clean as possible. Particulate matter released in the room by personnel, equipment, or materials is swept away by the moving air before it has an opportunity to settle onto or into the product. The cleanest area of a clean room is nearest the supply ducts, and personnel, equipment, and other sources of contamination must be positioned "downstream" of items to be protected from contamination. The velocities of air in a clean room are usually set at around 90 to 100 feet per minute (about 1 mph). Care must be taken that no furniture, supplies, and other bulky items are placed in front of the supply ducts, which would disrupt the movement of clean air into the clean room.

The movement of personnel and materials in and out of the clean facility must be carefully controlled to avoid additional contamination. Atmospheric pressure inside a clean room must be slightly higher than the pressure in adjacent, less clean areas so contamination is prevented from entering through opened doors or cracks and openings in walls and ceilings. A pressure differential of about 0.05 inches (water gauge) is usually sufficient. Air locks with interlocking doors at major entrances are useful in stabilizing air pressure in the clean room, but the air supply to the clean room must be sufficiently robust to maintain the pressure differential whenever a door is opened. Extensive use of observation windows and intercommunication systems should be used to reduce excessive traffic in and out of the clean room by supervisory personnel and visitors.

Dressing rooms for cleanroom personnel must be immediately adjacent to the clean rooms. Various devices such as air showers and tacky floor mats are used to reduce contamination adhering to workers' attire as they enter the clean room. Additional discussion regarding dressing room design will be found in the section on gowning.

HEPA and ULPA Filters

The secret of maintaining ultraclean air is the passage of the supply air through HEPA (High Efficiency Particulate Air) filters, discussed in chapter 3, and ULPA (Ultra Low Penetration Air) filters. The pleated design of the HEPA filters allows for large airflows with greater than 99.99 percent filtration efficiency for particles 0.3 μm or larger with only a 1–2 inch pressure drop. Because most microorganisms are larger than 0.3 μm, HEPA filters are eminently successful in controlling microbial contamination. For airborne microorganisms less than 0.3 μm, particularly viruses, it is doubtful that they would survive long without the protective effect of larger particulates and, thus, do not generally appear to pose a problem of becoming a significant source of viable contamination. Because HEPA filters are not 100 percent effective, they cannot be considered as producing truly sterile air.

The ULPA filters use a polymer membrane as the filtering medium and exhibit efficiencies as high as 99.9999 percent for particles as small as 0.1 μm.

A typical HEPA filter unit (Figure 3.8) 2 $\times$ 2 feet (4 ft^2), and 9 inches deep, actually has over 100 ft^2 of filter area. The filter medium is pleated, the pleats are held apart with aluminum spacers, and the entire assembly is sealed in a metal or composition frame. The completed unit is installed with gaskets to prevent air leaks around the filter box.

Once installed, the filters must be thoroughly tested for leaks, for filters may experience damage in shipping or installation. Testing is done by scanning the entire face of each filter and surrounding framework with the probe of a particle counter while particles, such as dioctylphthalate (DOP) smoke, are generated in the supply duct. Particle counts on the exit side of the filters should not exceed 0.01 percent of the counts on the supply side. Periodic testing to confirm filter system integrity, as part of a certification program, should be carried out every 6–12 months to assure that the filters are functioning properly. Microbiological testing will be covered in chapter 5.

Aseptic Filling Areas

The final packaging of sterile pharmaceutical products can be handled in one of two principal ways. In TERMINAL STERILIZATION the product is placed into clean containers, sealed, and subjected to sterilization by heat or radiation. Note that product and container are not initially sterile but are handled in a manner that assures a low bioburden. ASEPTIC FILLING involves the placing of sterile product into sterile containers and sealing with sterile enclosures. All equipment that comes into direct contact with the product and the interior of the containers, including any gases, must be sterile and, when appropriate, be pyrogen free as well.

Environmental controls for facilities where aseptic filling is carried out must be considerably more stringent than what is needed for terminally sterilized products. All aseptic filling areas where sterile products and containers are exposed to the environment are called CRITICAL AREAS. According to FDA guidelines, critical areas must generally conform to the specifications of a Class M3.5 clean room plus exhibit a microbiological air quality of not more than 2.7 colony-forming units per cubic meter. This area is usually protected by a steady flow of HEPA-filtered air aimed directly at the work space. Human penetration into the critical area is kept to an absolute minimum and is done only with sterile tools or gloves in a manner that does not disrupt the protection of the unidirectional air flow. Workers must avoid reaching over critical areas; although air flow should sweep away most particles that fall from a worker's arm, larger particles may still fall into the product and contaminate it.

Surrounding the critical area is the ASEPTIC AREA, a slightly less clean (Class M4.5 to M5.5) but still highly controlled space. Full-barrier gowning is required for all personnel in this area, and all material and equipment must be either sterilized or thoroughly disinfected. Full-barrier gowns cover all exposed parts of the body, including the face, and are designed for minimum leakage around the neck, wrists, and ankles.

CONTROLLED AREAS are manufacturing and storage areas where nonsterile product, containers and other objects that may come into contact with the product are handled. Generally, these areas must conform to Class M6.5 specifications and show average microbiological air counts of not more than 67.5 colony-forming units per cubic meter.

Biohazard Cabinets

Biohazard cabinets are devices similar to chemical hoods but they are primarily used either to protect the worker and the surrounding facility from contamination by hazardous microorganisms being handled within the cabinet, to protect items in the work area from contamination from the surroundings, or both. Biohazard cabinets should not take the place of chemical hoods for handling hazardous chemicals because they cannot retain vapors and gases.

Three classes of biohazard or biological safety cabinet are available, and each class fulfills a particular purpose.

Class I

A Class I cabinet (Figure 4.2) that can be operated either with an open front or with a closed front through which arm-length gloves pass. Negative pressure maintained within the cabinet assures an inward flow of air that protects the worker and the surrounding area from hazardous viable aerosols that might be generated in the cabinet. When operated with an open face, velocity of entering air must be at least 75 feet per minute. All exhaust air is filtered through HEPA filters before being released into the atmosphere; some models also subject the exhaust air to ultraviolet light or heat decontamination before release.

Class II

A Class II cabinet (Figure 4.2) recycles a portion of the air that has been drawn into it, filters it, and redirects it over the work area. In this cabinet, not only is the worker protected by the inward flow of air, as in the Class I cabinet, but the work area is protected from outside contamination by the shower of filtered air. Thus, Class II cabinets have the distinct advantage of protecting both operator and product and are widely found in numerous research and industrial applications. However, neither Class I nor II cabinets are sufficiently effective to use with extremely hazardous organisms. For these, Class III cabinets must be used.

Class III

A Class III biosafety cabinet offers maximum operator safety. It is airtight, and the operator must manipulate objects in the cabinet through glove ports or by various remote control devices. These cabinets are usually reserved for handling organisms of the highest hazard level. All operations, such as inoculations, incubation, centrifugation, staining, and so forth, are done inside the cabinet or, occasionally, in several cabinets connected by airtight passthroughs. Air pressure in the cabinet must be maintained at least -0.5 inches (water gauge) relative to the surrounding area,

Figure 4.2. Two Types of Biohazard Cabinets. A Class I cabinet (top) where all of the exhaust air is HEPA filtered and discharged. While the worker is protected from contamination escaping the work area, no contamination protection is afforded the work area. A Class II cabinet (bottom) where a portion of the exhaust air is filtered and recirculated back into the cabinet to protect the work area from contamination. As with the Class I cabinet, the worker is protected from contamination exiting the work area as well.

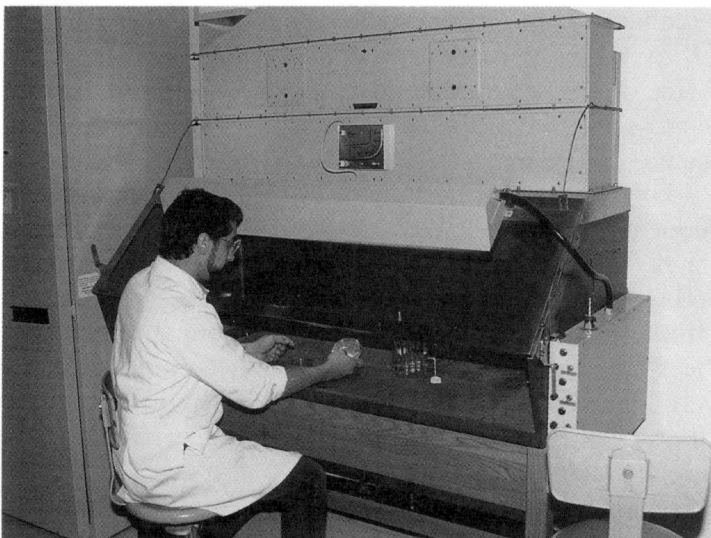

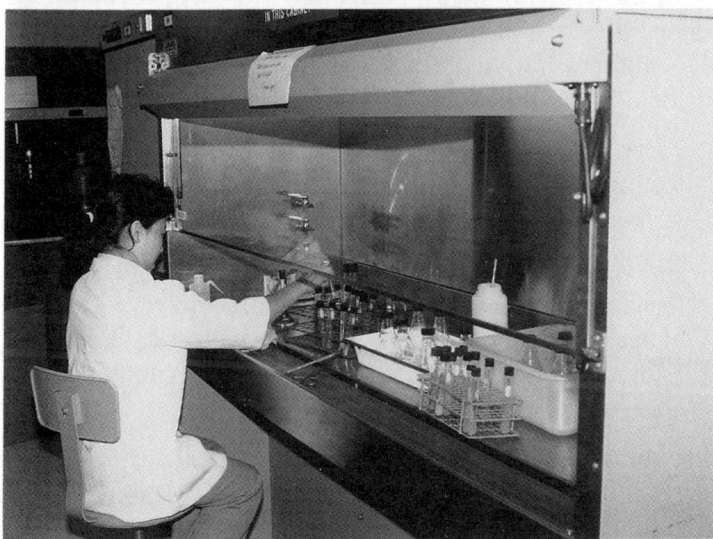

ensuring that no infectious material escapes in the event of a leak. Both supply and exhaust air are HEPA filtered, and the exhaust air is double filtered for additional safety.

The protection of laboratory workers or products afforded by Class II or open Class I biological safety cabinets is only as good as the care and skill of the workers who use them. Partially blocked openings or quick movements by workers within or near the opening can disrupt the inward airflow, allowing potentially contaminated air to enter or escape the cabinet. For example, because the inward air flow is usually less than 100 feet per minute (slightly over 1 mph), a person walking past the cabinet at a moderate pace of, say, 2 mph would easily create a disruptive shock wave across the face of the cabinet that could reduce the protective feature of the inward flow of air. Any cross draft from an open door or a nearby heating or air-conditioning vent would also reduce the efficiency and safety of a biohazard cabinet. The placement of biohazard cabinets must therefore be carefully planned to avoid exposure to foot traffic and excessive air movement from open doors or heating/air-conditioning vents.

The working area of a biohazard cabinet should not be excessively filled with bulky equipment, such as water baths and centrifuges, that blocks the normal flow of air exiting at the rear or along the edge of the opening.

Biological safety cabinets should be certified when first installed and at least annually thereafter plus whenever they are moved or filters are changed. Certification should be conducted with the cabinet's normal complement of equipment in place and should include the following tests:

- Filter leaks—Using a particle counter as described for the testing of cleanroom filters, the entire filter faces plus all seals should be probed.

- Air velocity—Air coming out of the filter bank and air flowing into the front opening should have a velocity of at least 75 feet per minute as measured by a handheld anemometer. The entire filter face and cabinet opening should be velocity tested.

- Cabinet integrity—Sheet metal fatigue may cause seams to open. Such defects can be detected by sealing the front opening of the cabinet and the exhaust vent with heavy plastic sheeting and injecting a halocarbon into the cabinet to a pressure of 0.5 inches. Using a halogen sniffer, any leaks in the cabinet's metalwork are quickly discovered and can be sealed with silicon.

Microbiological Clean Benches

Workers must avoid confusing clean benches with biohazard cabinets. Clean benches (sometimes called pharmacy benches) appear superficially like biosafety cabinets, but the principal function of clean benches is to protect media or other items in the hood from contamination by an outward flow of HEPA-filtered air. They offer no protection for the worker and, in fact, are hazardous if used to handle harmful chemicals or pathogenic microorganisms, for the air moves from the interior of the hood into the face of the worker (Figure 4.3).

Biohazard Levels

In 1976, the National Institutes of Health (NIH) developed guidelines to be used in laboratories that engage in research involving recombinant DNA (genetic engineering). Included in the guidelines are definitions of four classes of laboratory facility, ranked according to the relative levels of protection afforded workers conducting experiments in the laboratories. Experiments that constitute little or no health risks to humans can be carried out in a P1 facility, which has a minimal level of safety features, whereas experiments of greatest risk require a P4 facility. Since the publication of the NIH guidelines, the Centers for Disease Control (CDC) have also developed similar guidelines for handling pathogenic microorganisms, adopting the facility definitions of the NIH document but renaming them BL (for Biohazard Level) 1–4. Thus, organisms not known to be hazards to human health may be handled in a BL1 lab while most known human pathogens are to be handled in either BL2 or BL3 labs. BL4 facilities are reserved for the most hazardous pathogens. Table 4.2 summarizes some of the principal features of the various biohazard levels.

Cleanroom Garments

The human body is a rich source of viable and nonviable particles that could contaminate a clean room. One solution to avoid this problem is to garb cleanroom workers in airtight suits that prevent all particles from escaping. This would work perfectly well if cleanroom personnel were reptiles, but that is not the case. In carrying out normal cleanroom operations, workers generate a considerable amount of body heat that must be dissipated through perspiring. Some attempts have been made to use air-tight uniforms with air conditioning that is either supplied by backpack units or delivered by hoses. This approach is used in BL4 laboratories and may be satisfactory in industrial applications so long as the workers are not required to move about to any great extent. Although newer fabrics that "breathe" but prevent particulate penetration have recently been developed for clean-room garments, leakage can still occur through openings at the neck and wrists if they are not properly sealed. In addition, an uncovered face releases a considerable amount of particulate

Figure 4.3. A Bacteriological Laminar Flow Bench. These should only be used to protect product from contamination but never to handle hazardous organisms. Arrows show direction of airflow.

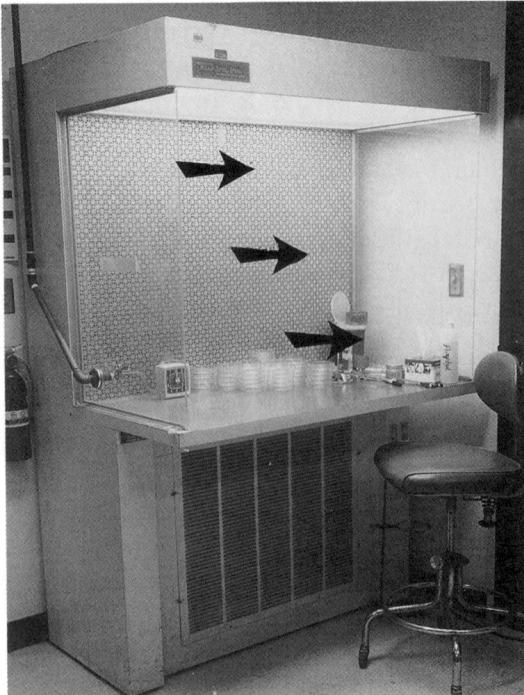

TABLE 4.2.
PRINCIPAL FEATURES OF BIOHAZARD (BL) LEVELS

BL1   1. Designed for easy cleaning.

        2. Bench tops impervious to water and resistant to acids, alkalies, organic solvents, and moderate heat.

        3. Lab furniture sturdy and spaced for easy cleaning.

        4. Hand washing sink available in laboratory.

        5. Windows that open have fly screens.

BL2   1–5 above, plus

        6. Autoclave available for sterilization.

        7. Personnel have special training to handle pathogens.

        8. Controlled access to facility.

        9. All procedures that produce aerosols carried out in biohazard cabinets.

BL3   1–4 and 6–9 above, plus

      10. Access to facility through double set of self-closing doors.

      11. Windows are closed and sealed.

      12. Ducted ventilation system draws air into facility from entry area; exhaust air is not recirculated but is exhausted to outside and away from occupied areas and intake ducts.

BL4   1–4 and 6-12 above plus

      13. Maximum containment design and equipment.

      14. Personnel enter and leave only through clothing changing and shower rooms.

      15. All equipment, materials, waste, and other items must be decontaminated prior to leaving facility.

      16. All exhaust air is HEPA filtered.

      17. All work is conducted within Class III biohazard cabinets by personnel in one-piece positive pressure suits.

From CDC, National Institutes of Health. 1993. *Biosafety in microbiological and biomedical laboratories.* 3rd ed. DHHS Publication no. (CDC)93-8395. Washington, DC: U.S. Department of Health and Human Services, Public Health Service.

contamination from the mouth, nose, and eyes and from the sloughing of skin particles (Table 4.3).

Full-barrier suits are frequently required where particulate release from exposed areas of the face must be prevented. These suits cover the entire face with mask and goggles or a space helmet. The helmets may have various types of one-way breathing valves that allow air to enter the helmet, but close when the wearer exhales. Expired air is thus forced down the neck and into the bodysuit, where it filters through the suit material.

Correct Gowning Procedures

Clothing worn under cleanroom garments should be clean, non-frayed, and nonlinting. Gowning is conducted from the head down. This is to prevent contamination from dropping from street clothes onto cleanroom garments. The commonly accepted method for proper gowning is as follows:

1. All excess personal property (coats, umbrellas, sweaters, and hats) should be stored outside the gowning room, preferably in a separate locker room. Unnecessary personal items of value, such as jewelry, should be left at home; watches and wallets may be taken into the clean room provided they are kept in inner pockets and not taken out. Other items such as keys, pens, and combs, should be left in the locker room. If required by company

Table 4.3.
PARTICLES RELEASED DURING VARIOUS ACTIVITIES

Activity	Particles Released/Minute 0.3 μm and Larger
Motionless: Sitting or standing	100,000
Walking about 2 mph	5,000,000
Walking about 3.5 mph	7,500,000
Walking about 5 mph	10,000,000

From: Fitch, H. D. 1990. The human factor in cleanrooms: A host of problems. *CleanRooms* 4 (9):14–16. Used with permission.

policy, street clothing is removed before entering the gowning room.

2. All makeup must be removed, or better, not used at all. This should be done in the locker room. Approved moisterizing lotions may be applied to face and hands to reduce skin flaking.

3. All gowning items, including jumpsuits, smocks, booties, gloves, hoods, and masks should be left in their clean packaging or storage hooks until needed. Carefully inspect them for frayed edges, open seams, tears, or missing or broken zippers or snap fasteners. Such defective items must not be used, but discarded or set aside for repairs. Workers should select garment pieces carefully, especially hoods and jumpsuits, for proper size. Items too small cause strains at openings and leakage; garments that are too large can create a bagpipe effect, forcing contaminated air to exit the garment around sleeves and neck openings.

4. The gowning room floor should be clearly marked as to a "clean" side and a "dirty" side. Under no circumstances should personnel walk in the clean area in street shoes. Most facilities require workers to remove street shoes or don shoe covers. Putting on head covers and donning gloves prior to entering the gowning room may also be required to protect clean garments from becoming contaminated while being handled.

5. Ideally, a low bench should be installed along the line dividing the clean side from the dirty side.

6. Hair covering (bouffant), hood, and face coverings are put on first.

7. The coverall is donned by holding the left and right legs at the cuff, ankle, and midsection with left and right hands, respectively, and stepping into each leg one at a time, taking care not to drag any part of the coverall on the floor or bench.

8. Once the coverall is donned and while sitting on the bench, one foot is moved to the clean side of the line and without it touching the clean floor, its bootie is put on.

While keeping the foot with bootie on the clean side, the other foot is moved to the clean side and its bootie is donned.

9. Finally, gowning gloves are removed and one or two layers of work gloves are put on. If necessary to avoid air leakage, particle-free tape can be used to seal the junctures where gloves overlap the sleeves.

Gowned personnel entering a clean room should inspect each other for such things as exposed hair, skin, and open snaps. In addition, full-length mirrors should be placed at all cleanroom entrances for self-inspection.

Normally, regowning is necessary whenever personnel leave the clean room to enter dirtier areas and return. Company policy will dictate whether personnel must change to fresh gowns when returning. If gowns are to be reused, they must be carefully hung up in a special location in the gowning area where they cannot become contaminated. Gowns must be laundered and packaged in clean facilities that are at the same level or better than that in which they are used.

Sterile Gowning

Certain types of products require special protection during their manufacture, such as those packaged by aseptic filling. In these and other circumstances the use of sterile gowns, masks, hoods, and gloves by workers may be required. Generally, they are donned in the same manner as are nonsterile cleanroom garments with the following exceptions: personnel scrub their hands and arms with antimicrobial soap, don sterile gloves, and decontaminate their gloves with 70 percent isopropanol frequently during the gowning process.

Personnel Practices

While physical facilities and worker garments play critical roles in reducing contamination, good personnel practices become a third important factor in maintaining low contamination levels. These practices must be followed by everyone entering a clean room, including the normal cleanroom workers, inspectors, maintenance personnel, cleaning crews, visitors, even the company president.

Good personal cleanroom practices are not inherited, but must be acquired through comprehensive and conscientious TRAINING. All employees who regularly enter clean rooms, including maintenance and supervisory personnel, must complete a training program that includes at a minimum:

- Fundamentals of cleanroom design, operation, and monitoring

- Personal hygiene

- Proper behavior in the clean room

- Correct gowning procedures

- General microbiological principles, including sterility and aseptic techniques (if appropriate)

- Types of products handled in the clean room

- Safety

In addition, summaries of company policies regarding cleanroom operations, gowning, and other important information should be conspicuously posted in all cleanroom dressing rooms.

Personal Hygiene

Proper cleanroom behavior begins at home. Scrupulous care of the skin and hair must be practiced on a regular basis. Frequent bathing and shampooing will assure minimal shedding of skin fragments. Personnel with temporary skin conditions such as severe sunburn or those with colds or allergies that cause abnormal nasal flow, coughing, or sneezing should be assigned to non-cleanroom areas until the condition passes. Cosmetics, including nail polish, hair sprays, and skin medications, must never be worn in the clean room. Men should be clean shaven at all times or wear appropriate masks.

General Practices

The following discussion covers numerous general practices that have been shown to improve the effectiveness of the clean air features of a clean room for both viable and nonviable particulate contamination.

Avoid all rapid movements in the clean room. These not only disrupt the protective action of unidirectional airflow, especially around biohazard cabinets, but they also cause increased leakage of particles at the neck and wrists by a bagpipe effect. Nervous movements like head scratching or chin rubbing may help one think, but they can contaminate the worker's glove and generate additional airborne particulate contamination. They should be avoided. One should never open and reach into his or her garment to retrieve an item while in a clean room. Talking also contributes to contamination levels and should be limited.

Personnel should not smoke immediately before entering a clean room as smokers have been shown to continue to exhale thousands of smoke particles long after their last puff.

Maintaining tools and materials in a neat and orderly fashion makes it easier to keep work areas clean and also reduces unnecessary movements. Tools and other objects that fall to the floor should never be used until they have been cleaned. To recover an object on the floor, pick it up with a second tool and then clean both items. Alternatively, pick the object up with a wipe cloth, clean the object, and discard the wipe cloth. Do not allow your hand or glove to touch the floor or the dropped item. Gloves should be changed whenever they are contaminated. Gloves should also be replaced immediately whenever they sustain rips or holes.

Products that fall to the floor must not be returned to the process stream until they are inspected and cleaned, or discarded per company policy. It has been estimated that nearly 75 percent of pharmaceutical product failures because of contamination can be traced to the improper handling of products and containers by personnel.

Never bring any paper products, such as books, notepads, or wrappers into a clean room unless they are specially designated for cleanroom use. Use only approved writing instruments, never pencils. Ordinary paper and pencils are known to shed large numbers of particles.

Finally, eating and drinking, including gum-chewing, are never permitted in a clean room. Not only are many food products and their wrappers considerable sources of particulate contamination, movements of the mouth and face during eating will also contribute great numbers of particles to the surrounding air.

Summary

Many important factors impact the air quality of a clean room: the physical facilities, incoming raw materials, and the personnel who operate in the facilities. Strict manufacturing standards and a thorough testing program can prevent problems caused by dirty raw materials. Through the use of HEPA filtration, air entering a clean room is exceptionally free of particulate contamination. Personnel working in the clean room become a potentially significant source of contamination. The contribution that personnel make can be minimized by proper training in the use of garments and tools, and careful personal practices both at work and at home.

5 Detection and Enumeration of Microorganisms in the Clean Room

It is frequently necessary to assess the microbial levels in the air and on the surfaces of clean rooms in research and manufacturing facilities in order to confirm that biocontamination controls are operating properly. In addition microbial levels of process water and gases, raw materials, or final products often must also be determined, as well as the microbial levels of the personnel themselves who are working in the facility.

Airborne microorganisms are generally collected by drawing room air through a sampling device that traps particulates either in a liquid or on an agar surface. If collection is in a liquid, aliquots of the liquid can then be assayed for viable microorganisms by routine plating methods. If agar is used directly, and the agar is supportive of the growth of the microorganisms of interest, it need only be incubated under appropriate conditions to produce visible colonies that can be counted and related back to the volume of air collected.

Surface contamination can be estimated by sampling a known area with cotton swabs moistened with sterile liquid and suspending the collected contamination in a liquid for subsequent assay. Surfaces can also be sampled with contact plates, which are small plastic dishes that contain a solid nutrient medium that is pressed against the surface to be assayed. Other varieties of slides or strips that are coated with agar medium are also used for sampling surfaces. In all cases, a certain proportion of the particulate contamination on the surface to be sampled is transferred to the agar where under appropriate culture conditions the formation of colonies occurs.

Liquids and gases are conveniently assayed for viable contamination by passing them through filters of porosities small enough to trap the microbes, after which the filters are cultured. This is accomplished by simply placing the filters onto an appropriate agar medium or onto filter paper saturated with sterile liquid medium. Nutrients saturate the membrane filter and colonies form directly on the filter surface.

To determine microbial contamination in solids they must be reduced to small particulates by grinding or crushing, or by dissolving them into an appropriate solvent in order to release entrapped microorganisms. Obviously, care must be practiced when mechanically reducing solids to particles small enough to release microorganisms, for that requires exposing the solids to significant stress, which might be fatal to the microorganisms. Generally, aqueous solvents are preferred to dissolve solids in which microorganisms are embedded. Nonaqueous (organic) solvents, such as alcohol, toluene, ether, or hexane, may kill the cells and should be avoided.

The Sterility Suite

A laboratory specifically designated for processing and analyzing the microbiological tests that are conducted in a clean facility is frequently referred to as a STERILITY SUITE. Obviously, the microbiological quality of the sterility suite must equal or exceed the strictest levels of the entire facility, for contamination of samples cannot be tolerated. Sterility tests of a product, for example, that become contaminated during processing produce false positive results that could create enormous problems.

Let us now look at some of the techniques used by the clean facility microbiologist to assess microbiological quality of cleanroom environments, materials, and personnel.

Air Samplers

One of the major routes of product contamination is the surrounding air. Two general types of samplers are used for assaying airborne microorganisms: IMPACTORS and IMPINGERS. Impactors collect airborne particulate contamination onto a solid or semisolid surface; impingers collect the particles in a liquid. As defined earlier (chapter 4), a VIABLE PARTICLE is one that has one or more living microorganisms attached to it and, if captured on a nutrient medium, results in the formation of a colony following suitable culture procedures.

Because a given airborne particle, like a speck of dust or a skin flake, nearly always has thousands or more viable microbial cells associated with it, the colonies that form from the viable particle do not necessarily reflect the true microbial population of the air that was sampled. When captured, the particles may break up to varying degrees depending on the type of sampler used. Consequently, airborne microbial populations are usually reported in terms of COLONY–FORMING UNITS (CFU) per unit volume of air, which avoids any implication regarding a precise population determination.

The Andersen Sampler

The Andersen sampler (Figure 5.1) (Graseby-Andersen, Atlanta, GA), developed in the 1950s by Dr. Arie Andersen, remains one of the most widely used viable air samplers. Its high efficiency coupled with its ability to fractionate collected particles into several size ranges is the major factor that contributes to its popularity. Various models of the Andersen sampler have eight, six, two, or one stages. The size ranges of collected particles for the six-stage model are shown in Table 5.1.

Each stage consists of a perforated plate that is positioned above an open dish of agar medium. The holes in each stage are the same size but become smaller in each successive stage that the air passes through. The smaller the hole, the greater is the velocity of the air that is directed onto the agar surface. Because of their greater inertia, larger particles are impacted in the upper stages while smaller particles pass through the upper stages and are collected in succeeding stages where the velocities of the air impacting the agar surfaces are greater. The sampler collects air at the rate of one cubic foot per minute. The resulting colonies that form are counted and reported in terms of CFU per cubic foot of air according to the size of the collected particles.

When sampling air that contains moderate to high levels of viable particles, the probability becomes quite high that two or more viable particles may impact the agar at or near the same

Figure 5.1. Andersen Viable Air Sampler. Sampler with pump (top) and cross section of sampler (bottom) showing the six stages that hold petri dishes with agar medium. (Cross-section diagram adapted from Gregory, P. H. *Outdoor Aerobiology*. Oxford Biology Reader #62. London: Oxford University Press.)

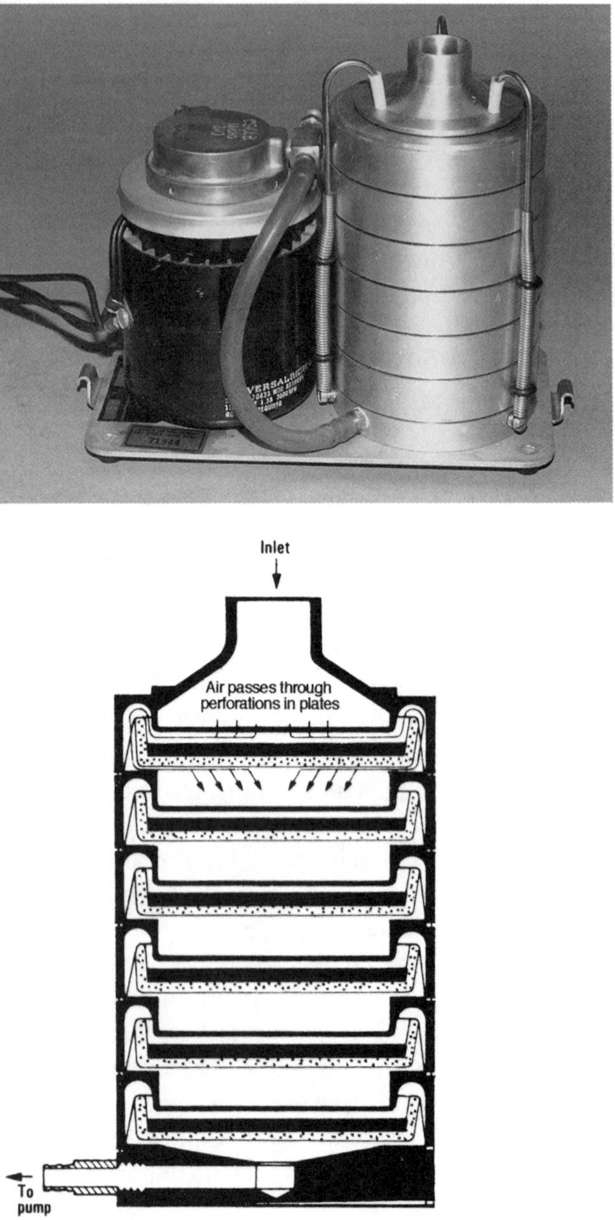

point to produce a single colony. The manufacturer of the sampler supplies a table to correct for this. Overcrowded plates are usually not a problem in cleanroom monitoring, however. More often than not, airborne microbial counts in clean rooms approach the lower limits of sensitivity of the Andersen sampler. As a general rule, as airborne sampling counts drop below 30 colonies, their statistical certainty becomes questionable.

FDA guidelines for aseptic fill facilities limit viable airborne particles to 0.1 CFU/ft^3. On a six-stage Andersen sampler operating for 20 minutes (20 ft^3), that would result in two colonies per six plates, far below the recommended statistical lower limit of 30 per plate. Why not sample for longer periods of time? One must remember that living organisms are being collected and if extended periods of sampling are used, the collected cells begin to die off as a result of the drying effects of the air as it passes over the agar surface. Twenty to 30 minutes is generally considered the maximum time period for operating an impactor like the Andersen sampler. Samplers with somewhat higher collection rates are available, and others are under development. One so-called high volume sampler will be described in a later section.

In order to maintain reproducible results from an impaction sampler, the distance from the perforated plate to the agar surface must be kept constant. This is accomplished by assuring that all sampling plates contain precise amounts of agar (i.e., to a constant depth). This requires the use of some type of automatic filling device when plates are prepared.

TABLE 5.1.
SIZE RANGES OF A 6-STAGE ANDERSEN AIR SAMPLER

Stage	Collection Size Range (μm)
1	larger than 7.0
2	4.7–7.0
3	3.3–4.7
4	2.1–3.3
5	1.1–2.1
6	*-1.1

*Lower limit of stage 6 is uncertain.

The Slit Sampler

The slit or slit-to-agar sampler (Figure 5.2) (New Brunswick Scientific Co., Edison, NJ) is an impactor-type sampler in which the agar plate rotates under the collection slit. Because the rotation of the plate is carried out by a clock mechanism and is constant, one rotation per six hours, for instance, we can obtain colony counts that can be related to time of collection. This ability becomes very useful when we are trying to determine a source of contamination. For example, if a careful record is made of all of the activities in a clean room during the period of sampling, any significant change in counts during the period can be related to specific activities, such as the switching off of antistatic devices, the opening of a door, or the centrifugation of a bacterial culture (Figure 5.2).

Some slit-to-agar samplers have some means of adjusting the distance from the slit opening to the surface of the agar. This feature avoids the need to fill the culture dish with precise amounts of agar to ensure a constant distance between slit and agar.

The Reuter Centrifugal Sampler (RCS)

Although it is an impaction sampler, the RCS sampler (Biotest Diagnostics Corp., Denville, NJ) draws air by a different principle than we have seen in the other impaction devices. In this sampler airborne particles are impacted onto the collection surface by being accelerated by the blades of a centrifugal air turbine (Figure 5.3). The collection surface consists of small segments of growth medium held in depressions in a narrow, flexible strip of plastic (Figure 5.3). The strip slides into the inner wall of the centrifugal turbine where it collects particles. After sampling is completed the strip is removed, its cover is replaced, and the strip is incubated appropriately. Care must be practiced to avoid contaminating the sampling strip while handling it.

The RCS sampler is the most portable of the commonly available viable air samplers, being only slightly larger and bulkier than a large flashlight. It is battery operated and self-contained with a reported flow rate of 1.4 cubic feet per minute. A built-in timer allows collection times of from 0.5–4.0 minutes.

High Volume Samplers

As indicated earlier, the statistical reliability of low colony counts can be somewhat offset by using so-called high volume samplers.

Figure 5.2. A Slit-to-Agar Microbial Air Sampler. Clockwork in the sampler (top) allows a 150 mm agar dish to rotate slowly under the intake slit creating a distribution of viable particles over the agar surface. Following the development of colonies, a plot of colony counts versus time can then be made. A typical plot (bottom) shows the fluctuations in airborne contamination that occur as a result of employee activity during a work shift in a manufacturing facility. Increased counts occurred at the start of the shift (8 A.M.), the morning rest break (10:15 A.M.), the lunch break (12 noon and return at 1 P.M.), the afternoon break (3 P.M.), and the close of the shift (5 P.M.). The cleaning crew began work about 6 P.M.

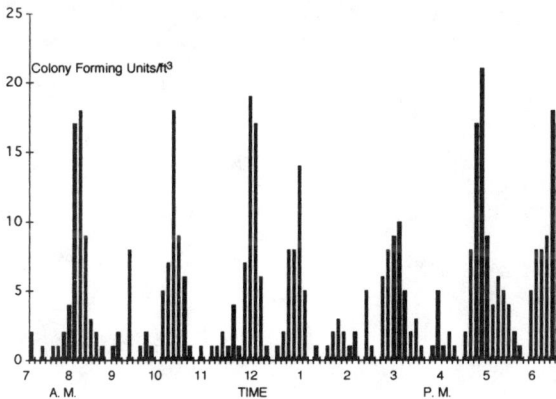

Figure 5.3. Air, Liquid, and Surface Samplers. An RCS centrifugal air sampler (top). Bottom, clockwise from top, agar collection strip for RCS air sampler, Hycor® Dip Slide for liquids, and Hycor® Contact Slide for surfaces. (Photos courtesy of Biotest Diagnostics Corporation, Denville, N.J.)

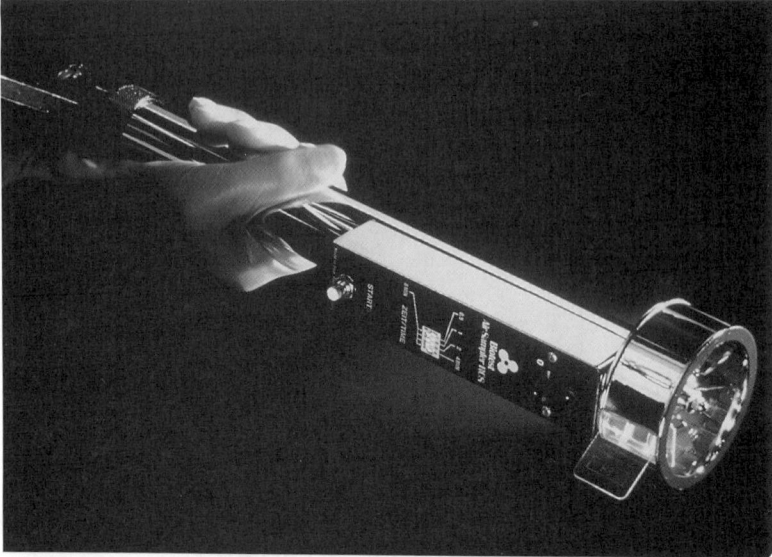

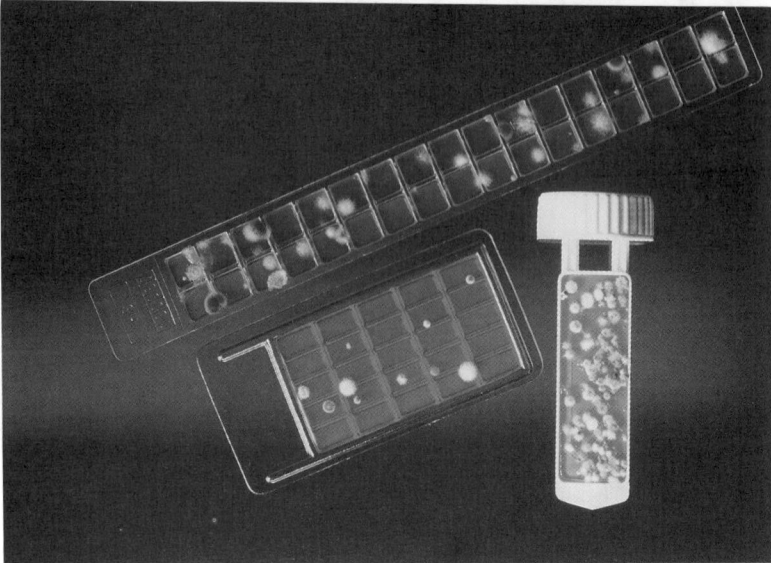

Whereas the more commonly used samplers collect air at rates of about 1 cubic foot per minute, the high volume samplers presently available, such as the SAS sampler (Spiral System Instruments, Inc., Bethesda, MD), collect air samples at a rate of 6.4 cubic feet per minute. At 1 colony-forming unit per cubic foot, a 20-minute sampling would result in about 128 colonies, which would be well within the acceptable range of the 30–300 colonies per plate rule. Samplers with even greater sampling rates are under development.

Liquid Impingers

While not widely used, impingers have one advantage not enjoyed by the impactors: collection liquids can be divided into several aliquots, each of which can be treated differently. The liquid impinger Model AGI-30 (Figure 5.4) (Ace Glass Co., Vineland, NJ) is filled with 20 ml of collection fluid. When the device is operated, a vacuum draws into the sampler a jet of air that is aimed at the bottom of the sampler. Airborne particles are impacted onto the bottom and are immediately suspended in the collection fluid. The sampler operates at 0.5 ft³/min. Following the collection period, the inner wall of the curved neck must be rinsed with sterile collection fluid, which is combined with the fluid in the sampler. This is necessary because to simulate the human nasopharynx, the neck was designed to trap particles larger than about 5 μm. The collection fluid can then be filtered through an appropriate membrane filter or plated on solid media in order to determine a viable count. Various aliquots may be plated onto media specific for bacteria, fungi, coliforms, or heat shocked at 80°C for 10 minutes to detect bacterial spores.

Sampler Efficiencies

No air sampler is 100 percent efficient, and samplers vary widely in sampling efficiency. Sampling efficiency is defined as the proportion of airborne particles that is actually trapped by a sampler. It is meaningless, for example, to compare results from one sampling site done with an Andersen sampler with those of another site done with an RCS sampler unless the two samplers were previously calibrated by running them side by side. Collection efficiencies of the same sampler can also vary depending on

Figure 5.4. An All Glass Impinger (AGI). Air is drawn into the curved opening at the top by a vacuum applied at the side arm. Airborne particles are trapped in the fluid at the bottom of the sampler after which the fluid is assayed for viable organisms.

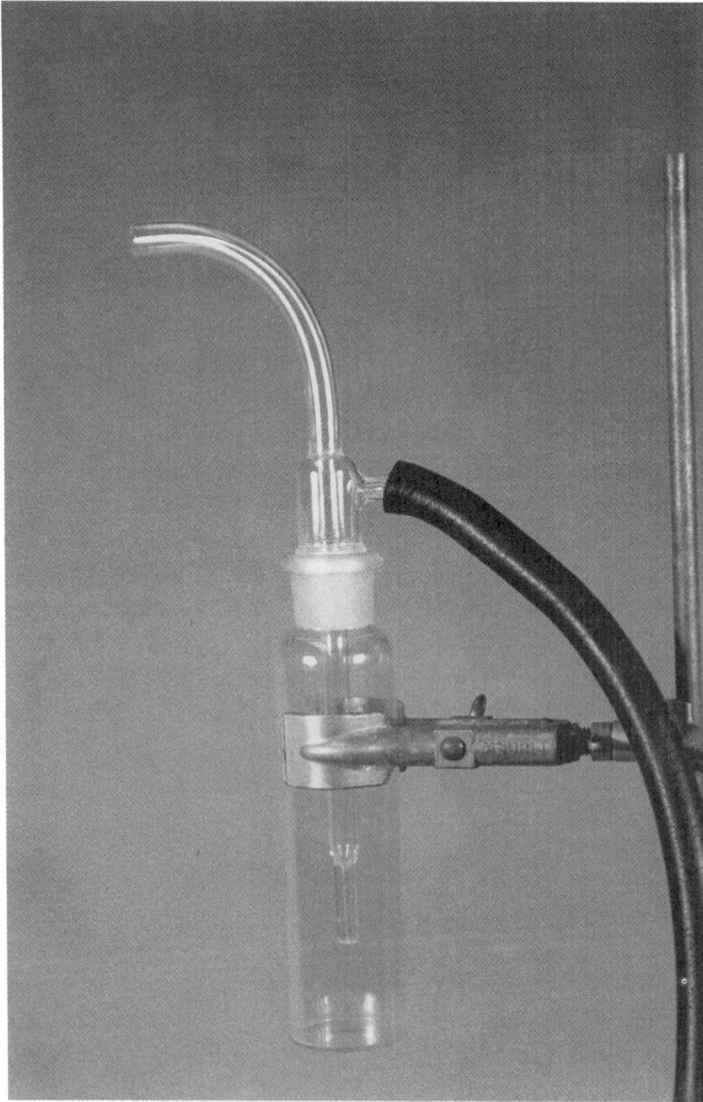

conditions, the most significant of which is the speed and direction of the room air relative to the sampler intake at the time of sampling. Put another way, the efficiency of a sampler in a horizontal-flow clean room may be different from that of the same sampler in a vertical-flow room unless the position of the sampler is adjusted so the two samplings are equivalent relative to air flow direction and velocity. For best results devices should be positioned relative to room air movement such that the sampling approaches isokinetic sampling. ISOKINETIC SAMPLING means the direction and velocity of the air entering the sampler is the same as the direction and velocity of the room air.

Fallout Methods

Although fallout methods do not necessarily give an accurate assessment of airborne contamination levels and should never take the place of an air sampling program, they can be a simple and relatively effortless measure of the general condition of a clean room. They are also useful in detecting specific areas of a clean room where airflow may fail to protect objects from biocontamination. There may be spots around equipment, for example, where the airflow is sharply deflected, causing particulates to deposit onto a surface of the equipment, or in or on the product or container. Localized areas where airflow velocities are reduced because of turbulence may also experience particulate fall-out. FALLOUT PLATES are simply culture dishes of appropriate growth agar that are left uncovered for varying lengths of time (usually an hour or more) to collect viable particles. The results of fallout plates are frequently reported in terms of colony-forming units per minute per square inch of agar.

FALLOUT or WITNESS STRIPS are small, sterile pieces of glass, metal or plastic that are placed uncovered about the clean room. Glass microscope slides can be used for this purpose. After appropriate time intervals, the strips are aseptically collected and placed into widemouth bottles of sterile 1 percent peptone water, which is then assayed onto suitable growth media. Some laboratories subject the bottles to agitation on a mechanical shaker or insonation in an ultrasonic bath before plating to ensure release of particles from the strips. If sample bottles are subjected to insonation, the bath liquid should be at least as high as the rinse liquid within the bottles and treatment should be carried out for at least 10 minutes.

In a manner similar to that described for handling liquid impinger collection fluids, the rinse liquid from fallout strips may be split several ways, allowing one to test, for example, for aerobes, anaerobes, and fungi all from the same fallout strip. One aliquot may be heat shocked at 80°C for 10 minutes to assess the presence of bacterial spores. The strips can also be cultured directly by placing them into sterile petri dishes and covering them with an appropriate molten agar growth medium. Following incubation, submerged colonies can be observed on the strips and counted.

Surface Sampling

It is required by government guidelines that all environmental surfaces in an aseptic fill area be monitored microbiologically on a regular basis to confirm that all contamination control precautions are operating satisfactorily. This can be accomplished by directly sampling surfaces such as floors, walls, and equipment parts, as well as workers' gowns and fingertips, with swabs or contact plates.

COTTON SWABS are especially useful for sampling irregularly shaped surfaces to obtain a general estimate of surface contamination. Sterile cotton swabs are moistened with sterile water or 1 percent peptone water and rubbed over a specified area of the object to be tested, say 4 square inches. The area should be covered at least 3 times, reversing the direction of rubbing between each time but using the same swab. Using aseptic technique, the head of the swab is then broken off and allowed to fall into a tube of 1 percent peptone water that is then shaken or subjected to insonation for 10–12 minutes. If contamination levels are expected to be high, dilution of the peptone water may be necessary before plating. If swabs are moistened with peptone water, the areas that are sampled must be cleaned after sampling with alcohol or detergent wipes to remove residual matter.

CONTACT PLATES, such as RODAC PLATES (Baxter Diagnostics, Inc., Scientific Products Division, McGaw Park, IL), are small plastic dishes that are overfilled with growth medium so that the level of agar rises above the rim of the plate (Figure 5.5). To sample a flat surface such as a floor, table top, or wall, the agar surface is firmly pressed against the flat surface using a gentle rocking motion to assure complete contact. Nonrigid surfaces, such as gowns or packaging material, can also be sampled in this manner if backed by a solid surface. The plates are then incubated under appropriate conditions, and the resulting colonies are counted. Results can be reported either in terms of CFUs per plate, or CFUs per square inch by dividing the CFUs per plate by 4 square inches. Again, it is recommended that nonpermeable surfaces that have been sampled by RODAC plates be cleaned with an alcohol or detergent wipe to remove residual medium transferred to the surface. Prepoured RODAC plates are commercially available with a variety of media. They should be used fresh, for after prolonged storage the agar surface shrinks into the dish making effective sampling difficult if not impossible.

Curved surfaces such as pipes, tanks, and filling equipment can be sampled with Hycor contact slides (Figure 5.3) (Biotest Diagnostics Corp., Denville, NJ), plastic paddles that are filled with appropriate agar medium. Because they are flexible and have a hinged handle, the paddles can be pressed against irregular or curved surfaces that are to be assayed for viable contamination. The strips are available with a variety of media with the option of having two different media on a single paddle.

If the surfaces to be sampled have residual disinfectant on them, the medium used in the contact plates should contain 0.5 percent Tween 80 and 0.07 percent lecithin as neutralizers. Each neutralizer must be sterilized separately and aseptically added to the medium just before preparing the plates. The need

Figure 5.5. A RODAC Plate. The plate is overfilled with melted agar. Once hardened, the agar can be pressed against any firm surface to estimate its level of microbial contamination.

Cover

Agar

Bottom

for neutralizers cannot be overstated, for if traces of disinfectants are transferred to the medium, bacteristasis may occur, limiting growth of the captured bacteria and leading to an underestimation of contamination levels.

Touch Plates

To detect the presence of possible sources of contamination, the sterile-gloved finger tips of workers in aseptic fill areas should be monitored at periodic intervals with TOUCH PLATES. These are dishes of sterile agar on which personnel leave their fingerprints. The plates are then incubated in the normal manner.

Sterile Media Fill Tests

Aseptic or sterile filling is where sterile product is added to presterilized containers and sealed under stringent contamination controls. One important method used to assess the effectiveness of an aseptic filling operation is to substitute sterile liquid growth medium for the product. The containers are then sealed and incubated. Any viable contamination that finds its way into the filling operation will be evidenced by the development of microbial growth. Media fill tests must be carried out under circumstances that simulate actual production conditions, including durations of operations, number of personnel present. and environmental background. Ideally, the tests should be conducted under worst case conditions such as at the end of an overtime shift or during other stressful situations that might reveal a weak link in the process.

Microbiological Assessment of Liquids

Liquids with viscosities near that of water can be passed through sterile membrane filters that have porosities of 0.22 μm, ensuring that most bacteria and fungi will be trapped. Viruses generally will not be retained by these filters unless they are strongly attached to larger particles. Viscous liquids, such as syrups, that are water soluble can be diluted with sterile water before filtration. The filters are then treated as described in chapter 2. Microbial populations of a liquid can also be estimated by using Hycor dip slides (Biotest Diagnostics Corp., Denville, NJ), which are agar-covered paddles that are immersed into the liquid and then incubated. Instrumental methods for determining microbial populations in liquids have been covered in chapter 2.

Microbiological Assessment of Solids

If it is necessary to assess microbial populations that are embedded in solids such as foods or pharmaceutical products, the organisms must be freed before they can be cultured. If the solid is water soluble, such as an uncoated tablet, it may be dissolved in sterile water and the fluid analyzed as discussed above. If the solid contains ingredients inhibitory to microbial growth, the cells can be collected on a membrane filter, quickly rinsed with sterile water, resuspended, and plated. If the material is not water soluble, it is not likely there is a solvent that would dissolve the solid and be nontoxic to the microorganisms; the solid usually must be mechanically reduced to particles small enough to release the organisms. This can be done in a sterile ball mill or mechanical grinder. However, it is virtually impossible to do this without significant cell destruction and a resulting underestimate of the viable population. The extent of cell destruction can be determined by mixing known numbers of microbial cells into a solid formulation and then subjecting that to the isolation procedure.

Disposal of Environmental Cultures

All liquid and solid cultures from environmental samplings are classified as biomedical waste and should be treated as biohazardous material. That is, prior to discarding it must be sterilized in an autoclave certified for biomedical waste for at least one hour (as a safety margin), destroyed in an approved incinerator, or collected by a biohazardous waste disposal service company.

Alert and Action Levels

An environmental sampling program is useless unless appropriate action is taken when excessive contamination is detected. Maximum levels of contamination, which when exceeded, should trigger specific action that is defined by company policy. Such levels are generally conservative and based on long historical records of the facility. An ALERT LEVEL is a level of contamination that, if exceeded, alerts appropriate personnel that there is an apparent deviation from normal operating conditions. Action may or may not be necessary at this stage depending on company policy although retesting of the facility in question would seem prudent. A higher level of contamination, designated the ACTION LEVEL, if exceeded, is answered by appropriate action that would be directed toward discovering the nature and source of the contamination and controlling it. If necessary, more drastic action such as shutting down a filling operation and quarantining product may be required.

Summary

A principal purpose of a clean room in a biomedical research or manufacturing setting is to reduce levels of microbial contamination. This is achieved through a combination of architectural features and personnel practices. Because microorganisms are not visible to the naked eye, there are only two ways to determine if this purpose of the clean room is satisfied: product failure and microbiological monitoring. Obviously, the latter approach is the preferred one.

A variety of methods is available to the microbiologist to measure levels of microbial contamination in the clean room. Several should be used on a regular basis, and the results maintained and displayed in a manner that can quickly alert workers of breakdowns in the operation of the facility. Company policy should include action levels, which are contamination levels that when exceeded trigger specific action. Such action might involve increasing disinfection frequencies, tightening personnel practices, or recertification of filters and other contamination control equipment. Clearly, the microbiologist carries an enormously important burden in maintaining the success of the operation of the bioolean room, and, ultimately, the success of the company.

Glossary

Absolute Barrier—An enclosure or cabinet constructed so that no contamination can pass between its interior and the surrounding area. A Class III Biological Safety Cabinet is an example of an absolute barrier (*see* Partial Barrier).

Aerobic—Refers to the growth of microorganisms in the presence of free oxygen, usually at normal air concentration.

Aerobiology—An area of biology concerned with the collection, enumeration, and identification of airborne microorganisms.

Aerosol—Liquid or solid particles suspended in a gas, usually air. Fog is an example of a natural aerosol. Aerosols containing microorganisms are frequently generated in the laboratory through such activities as centrifugation, blending and pipetting. A common cause of laboratory infections.

Agar—A polysaccharide extracted from certain types of sea kelp that is used to solidify microbiological growth media or the microbiological growth medium itself, including nutrients, that has been solidified by the addition of agar.

AGI or All-Glass Impinger—An air sampler that captures airborne particles into a liquid.

Anaerobic—Refers to the growth of some microorganisms in the absence of free oxygen.

Andersen Sieve Sampler—A microbiological air sampler having up to eight stages stacked in series, each stage consisting of a perforated plate. A petri dish of agar growth medium is placed beneath each plate. Air that is drawn into the sampler impacts the agar of each stage. Air velocity increases at each succeeding stage due to progressively smaller holes causing suspended particles to be distributed among the plates according to size. Larger particles are trapped in the upper stages; the smaller, in the lower stages.

Anisokinetic Sampling—The taking of air samples under conditions in which the velocity and direction of the air entering the sampler are different from the velocity and direction of the room air. Where the velocity of the sampled air is greater than that of the room air, smaller particles are selectively collected over larger particles. If the velocity is less, larger particles are collected in place of smaller particles.

Antiseptic—Chemical product capable of destroying or inhibiting microorganisms and that is safe to use on the body. *See* Disinfectant.

Aseptic—Literally, "without infectious matter," but often erroneously used as a synonym for "sterile" (i.e., absence of all living organisms).

Aseptic Area—The space immediately surrounding the Critical Area of an aseptic fill operation. Must conform to at least Class M4.5 contamination level.

Aseptic Filling—Sterile containers are filled with sterile product and sealed with sterile closures. Frequently used with products that cannot tolerate terminal sterilization.

Aseptic Technique—All practices that avoid microbiological contamination, such as flaming the mouth of a vessel after opening it and the wearing of sterile gowns and gloves.

Autoclave—A device for sterilizing items with steam under pressure.

Bactericidal—Capable of killing vegetative bacterial cells, but not necessarily bacterial spores.

Bacteristatic—Capable of stopping the growth of vegetative bacterial cells, but not necessarily killing them. Growth usually resumes when the bacteristat is removed.

Bioburden—The level of microbiological contamination in or on an object or material immediately before sterilization.

Biocide—An agent that kills organisms.

Bioclean—Special attention paid to controlling microbiological contamination. Synonymous with biocontrol.

Biological Indicator—A sample of bacterial cells (usually spores) of known type and number that is exposed to a sterilization procedure in order to determine if the procedure is working. The sample is cultured following its exposure and the absence of growth shows that sterilization conditions were met.

Biosafety Cabinet—Containment device for protecting workers from hazardous microorganisms and, in some cases, protecting the work area from environmental contamination.

Biosafety Levels—Four biosafety levels have been defined by the Centers for Disease Control (CDC) and the National Institutes of Health (NIH). Level 1 is for handling organisms not known to cause disease in normal, healthy adult humans. Levels 2, 3, and 4 are designed for handling organisms of increasing hazard for humans.

Biotechnology—Generally, any commercial exploitation of a biological process. More often applies to applications of molecular biology or genetic engineering.

Broth—Liquid growth medium for microorganisms.

Centrifugal Air Sampler—An impaction-type microbiological sampler that captures viable particles by accelerating them into a nutrient surface through a centrifugal blower. The RCS or Reuter sampler is an example of a centrifugal sampler.

Colony—A discrete, visible accumulation of microbial growth on the surface of a solid culture medium.

Colony-Forming Unit—That which results in the formation of a colony of microbial growth on appropriate solid media. Assumed to originate from one cell but difficult to prove.

Commercially Sterile—*See* Sterile.

Critical Area—Space immediately surrounding an aseptic filling operation.

Critical Orifice—A small opening through which air passes at a velocity that is independent of the pressure difference between either side of the opening, provided the difference is a minimum value, usually 0.5 atmosphere. Certain air samplers, such as the all glass impinger, operate with a critical orifice to control the flow rate of air. Sampling flow rate is unaffected by fluctuations in the vacuum source, and there is, therefore, no need to calibrate the flow rate of the sampler.

Cytotoxicity—The degree of harm toward animal cells exhibited by a substance. Usually in reference to disinfectants, antiseptics, and other products that may come in contact with body tissue.

D-Value—The decimal reduction time. The time it takes for a physical or chemical agent to reduce a microbial population by 90 percent, or one log. Expressed in terms of time (minutes or hours) when dealing with heat, and in megarads when ionizing radiation is involved. The reciprocal of the D-value, $1/_D$, is referred to as the death rate constant, K.

Decontamination—The removal or destruction of living organisms to some lower population level, but not necessarily to zero (*See* Sterilization).

Disinfectant—A chemical agent capable of killing vegetative forms of disease-causing bacteria, fungi, and viruses, but not bacterial spores. Used on inanimate objects. *See* Antiseptic.

DOP (Dioctylphthalate)—A chemical capable of producing smoke of very uniform particle size. Frequently used for testing air filters. Chemical synonym is di(2-ethylhexyl)phthalate.

Droplet—An airborne particle consisting primarily of liquid. Saliva droplets are frequently generated from talking, eating, sneezing, or coughing. While most droplets settle to the floor almost immediately, many remain airborne, dry, and become very small Droplet Nuclei, which remain suspended for long periods of time and add significant numbers of microorganisms to the air.

Droplet Nucleus—An airborne particle that originated as a droplet but which has dried (*See* Droplet).

Dry Heat Sterilization—Thermal sterilization at relative humidities less than 100 percent. It is less efficient than moist heat where RH is at 100 percent, requiring longer times at higher temperatures.

Enzyme—An organic catalyst, usually protein, that makes it possible for a cell to carry out chemical reactions at moderate temperatures. A typical bacterial cell may have one to two thousand different enzymes.

Ethylene Oxide—Colorless gas used as a sterilant. It has great penetration power and is effective at room temperature although a certain level of moisture is required for optimum effectiveness. It is toxic, flammable, and a suspected carcinogen. Its flammability can be reduced by mixing it with carbon dioxide or a fluorocarbon.

Fomite—An inanimate object or material that acts as an intermediate carrier of microbiological contamination. Examples might be doorknobs, tools, surgical instruments, and the like.

Facultative—Refers to certain microorganisms that are capable of growing under varying environmental conditions. For example, facultative anaerobes can grow under both aerobic and anaerobic conditions (*See* Obligate).

Gnotobiotic—Literally *known biota*, but generally refers to the breeding and maintenance of germ-free animals.

Halophile—A microorganism that can grow in the presence of high concentrations of salts.

Heat Shocking—A procedure in bacteriological assays to eliminate vegetative cells but to allow spores to remain. Usually carried out at 80°C for 10–20 minutes.

HEPA—High Efficiency Particulate Air filter.

Impactor—A type of air sampler that collects airborne particles onto a solid medium.

Impinger—A type of air sampler that collects airborne particles into a liquid.

Infective or Infectious Particle—A Viable Particle that has one or more pathogenic (disease producing) organisms on it.

Inoculum—A small quantity of viable microorganisms that is transferred to sterile medium to initiate new growth.

Insonation—*See* Sonication.

In-Use Dilution Testing—The testing of disinfectants, sanitizers, and antiseptics at concentrations that match those actually used by the consumer. Synonymous with Use-Dilution Testing.

Isokinetic Sampling—The collection of air samples under conditions in which the velocity and direction of the air entering the sampler are the same as the velocity and direction of the room air (*See* Anisokinetic Sampling).

K-Value—*See* D-Value.

Laminar Air Flow (LAF)—Air movement in which all air molecules are moving at the same velocity and direction. Seldom achieved except in relatively small spaces because of natural turbulence and turbulence caused by equipment, furniture, and other objects in the room.

Level of Assurance—An extrapolation of a killing curve to below one viable organism. Can be used as a probability that the population still contains a viable cell. For example, a level of assurance of 10^{-6} means there is a one in a million chance the population has not been sterilized.

Lipopolysaccharide—A molecule consisting of a lipid (fat) and a polysaccharide, a long chain of sugars. Found in the outer membrane of gram negative bacteria.

Lipoprotein—A molecule consisting of a lipid (fat) and a protein.

Media Fill Test—A validation test for aseptic filling by using sterile growth medium in place of product and looking for growth in the containers following appropriate incubation.

Membrane Filter—A filter made from any of a number polymeric materials such as cellulose, polyethylene, or tetrafluoroethylene. Membrane filters can be manufactured with very narrow ranges of pore diameters making these filters very useful in the collection and sizing of microscopic and submicroscopic particles, and in the sterilization of liquids.

Mesophile—A microorganism that grows best between 25°C and 4°C.

Microaerophil—A microorganism that grows best in the presence of trace amounts of gaseous oxygen, in contrast to an Aerobe, which grows best in the presence of the normal level of oxygen in the air.

Micron—An obsolescent term referring to the unit of length, 10^{-6} meters, 10^{-3} millimeters, or $1/_{25400}$ inch. Replaced by micrometer (μm).

Moist Heat Sterilization—Sterilization by heat at 100 percent relative humidity.

MPN (Most Probable Number)—A statistical method for estimating very low bacterial populations in liquids by observing the proportions of aliquots from the sample that contain no viable organisms and applying a form of the Poisson Distribution. Frequently used to assess the bacteriological quality of environmental water sources.

Obligate—Refers to microorganisms that are incapable of growing under more than one environmental condition. For instance, obligate anaerobic bacteria can only grow in the complete absence of gaseous oxygen (*See* Facultative).

Opportunist—A microorganism capable of causing disease only under special circumstances, such as in the case of a person with a severely weakened immune system. That is, an organism that is normally harmless but which takes advantage of an opportunity.

Parasite—An organism that derives its nutrients from other living organisms (*See* Saprophyte).

Parenteral—Products administered to patients by routes other than by mouth, such as intravenous (IV) or intramuscular (IM).

Partial Barrier—An enclosure or cabinet that because it is open offers worker and product partial protection from contamination. A Class II Biological Safety Cabinet is an example of a partial barrier (*See* Absolute Barrier).

Pasteurization—A process that heats products, such as milk or beer, just sufficiently to kill pathogenic (in the case of milk) or spoilage (in the case of beer or fruit juices) organisms without significantly altering flavor. Milk is usually pasteurized by heating it to 161°F (72°C) for 15 seconds.

Pathogen—An organism capable of producing disease.

Peptone—Plant or animal protein that is partially digested with acid or enzymes. Used as a nutrient in microbiological growth media.

Phenol Coefficient—A measure of the effectiveness of a disinfectant by comparing it with phenol (carbolic acid). For example, a phenol coefficient of 100 means the disinfectant is equivalent to phenol in killing power when diluted to $1/100$ the concentration of the phenol. (*See* Use-Dilution Test.)

Phospholipid—A molecule consisting of a lipid (fat) and phosphorus. Common component of cell membranes.

Poisson Distribution—A statistical calculation that predicts how often an event will occur by considering how often it doesn't occur and how many opportunities is has to occur.

Polypeptide—A chain of amino acids linked together by peptide bonds.

Psychrophile—A microorganism that grows best at temperatures below 20°C.

Pyrogen—A substance that produces fever.

RCS—Reuter Centrifugal Sampler (*See* Centrifugal Air Sampler).

RODAC Plate—A culture plate designed so that when filled with sufficient agar the medium rises above the rim of the plate. The plate can then be used to measure the microbiological levels of smooth surfaces by pressing the raised agar against the surface, thereby transferring some of the organisms to the agar. Following suitable incubation, visible colonies appear on the medium indicating the level of contamination of the surface sampled.

Sanitizer—A chemical agent capable of reducing microbial contamination to levels as ordered by public health guidelines or regulations. Usually associated with operations that handle food and beverages such as food manufacturers, restaurants, and cocktail bars.

Saprophyte—A microorganism that derives its nutrients from nonliving organic matter (*See* Parasite).

Settling Plate—A method of assessing microbiological fallout by leaving uncovered petri dishes containing growth medium in selected locations for certain periods of time. Following appropriate incubation, the numbers of colonies that form on the agar are an indication of how effective the air movement in a bioclean facility is in preventing airborne contamination from settling onto work or product surfaces.

Sieve Sampler—*See* Andersen Sieve Sampler.

Slit Sampler (Slit-to-Agar Sampler)—An impaction sampler that collects airborne particles onto a slowly rotating agar plate, making it possible to obtain the microbiological profile of a facility as a function of time.

Sonication—A method of cleaning objects by subjecting them to high frequency sound while submerged in a liquid. Also used to disintegrate microbial cells for certain biochemical applications. Synonymous with Insonation.

Spore—A subcellular body that some species of bacteria form and that is considerably more resistant to harsh conditions, such as heat, disinfectants, and radiation, than the vegetative form. Bacterial spores are more correctly referred to as endospores to distinguish them from the spores formed by molds, which are somewhat less resistant to harsh conditions.

Sporicide—An agent capable of killing bacterial spores.

Sterile—Free of all living organisms.

Sterile, Commercially—Free of all living organisms that are pathogenic or will cause spoilage during the normal shelf life of a product. Usually applied to canned foods.

Sterilization—The complete removal or destruction of all living organisms.

Terminal Sterilization—A process where nonsterile containers are filled with nonsterile product in the normal manner, then sealed and sterilized as the final step.

Thermopile—A microorganism that grows best at temperatures over 40°C.

Use-Dilution Test—*See* In-Use Dilution Testing.

Validation—Full, detailed documentation that all processes and procedures are functioning in the manner they were designed for. Required by the FDA.

Vegetative Cell or Form—A microbial cell capable of division as opposed to the spore form, which cannot divide.

Viable—Capable of life. The ability of a microorganism to grow and form visible colonies on an appropriate medium.

Viable Particle—A particle, such as dust, lint, or skin cells, that has one or more viable microbial cells on it.

Viricide—An agent that kills viruses.

Recommended Readings

Chapter 1

Atlas, R. M. 1995. *Principles of microbiology.* St. Louis, MO: Mosby.

Black, J. G. 1993. *Microbiology: Principles and applications,* 2nd ed. Englewood Cliffs, NJ: Prentice Hall.

Dale, J. W. 1989. *Molecular genetics of bacteria.* New York: John Wiley and Sons.

Dulbecco, R., and H. S. Ginsberg. 1988. *Virology.* Philadelphia: J. B. Lippincott.

Jensen, M. M., and D. N. Wright. 1993. *Introduction to microbiology for the health sciences.* Englewood Cliffs, NJ: Prentice Hall.

Maloy, S., J. Cronan, Jr., and D. Freifelder. 1994. *Microbial genetics,* 2nd ed. Boston: Jones and Bartlett.

Pelczar, M. J., E. C. S. Chan, and N. R. Krieg. 1993. *Microbiology: Concepts and applications.* New York: McGraw-Hill, Inc.

Stanier, R. Y., J. L. Ingraham, M. L. Wheelis, and P. R. Painter. 1986. *The microbial world.* Englewood Cliffs, N.J: Prentice Hall.

Tortora, J. G., B. R. Funke, and C. L. Case. 1995. *Microbiology: An introduction,* 5th ed. Redwood City, CA: Benjamin/Cummings Publishing Co., Inc.

Wistreich, G. A., and M. Lechtman. 1988. *Microbiology,* 5th ed. New York: MacMillan.

Chapter 2

Carlberg, D. M. 1988. Determining the effects of antibiotics on bacterial growth by optical and electrical methods. In *Antibiotics in laboratory medicine,* edited by V. Lorian. Baltimore, MD: Williams and Wilkins.

Collins, C. H., and P. M. Lyne. 1970. *Microbiological methods.* London: Butterworths.

Difco Laboratories. 1984. *Difco manual of dehydrated culture media and reagents for microbiological and clinical laboratory procedures.* Detroit, MI: Difco Laboratories.

Gerhardt, P., ed. 1994. *Methods for general and molecular microbiology.* Washington, DC: American Society for Microbiology.

Meynell, G. C., and E. Meynell. 1965. *Theory and practice in experimental bacteriology.* Cambridge: Cambridge University Press.

Nelson, W. H., ed. 1985. *Instrumental methods for rapid microbiological analysis.* Deerfield Beach, FL: VCH Publishers.

Norris, J. R., and D. W. Ribbons, eds. 1969–. *Methods in microbiology* (Series). New York: Academic Press.

Chapter 3

Bader, F. G. 1986. Sterilization: Prevention of contamination. In *Manual of industrial microbiology and biotechnology,* edited by A. L. Demain, and N. A. Solomon. Washington, DC: American Society for Microbiology.

Banerjee, K., and P. N. Cheremisinoff. 1985. *Sterilization systems.* Lancaster, PA: Technomic Publishing Co., Inc.

Block, S. S. 1991. *Disinfection, sterilization, and preservation,* 4th ed. Philadelphia: Lea and Febiger.

Mittleman, M. W., and G. G. Geesey. 1987. *Biological fouling of industrial water systems: A problem solving approach.* San Diego, CA: Water Micro Associates.

Parenteral Drug Association, Inc. 1982. *Design concepts for the validation of a water for injection system.* Technical Report #4. Philadelphia: Parenteral Drug Association, Inc.

Parenteral Drug Association, Inc. 1981. *Validation of dry heat processes used for sterilization and depyrogenation.* Technical Monograph #3. Philadelphia: Parenteral Drug Association, Inc.

Parenteral Drug Association, Inc. 1981. *Validation of steam sterilization.* Technical Monograph #1. Philadelphia: Parenteral Drug Association, Inc.

Parenteral Drug Association, Inc. 1980. *Validation of aseptic filling for solution drug products.* Technical Monograph #2. Philadelphia: Parenteral Drug Association, Inc.

Phillips, G. B., and W. S. Miller, eds. 1973. *Industrial sterilization.* Durham, NC: Duke University Press.

Chapter 4

CDC, National Institutes of Health. 1993. *Biosafety in microbiological and biomedical laboratories,* 3rd ed. DHHS Publication no. (CDC)93-8395. Washington, DC: U.S. Department of Health and Human Services, Public Health Service.

Fleming, D. O., J. H. Richardson, J. J. Tulis, and D. Vesley, eds. 1994. *Laboratory safety: Principles and practices,* 2nd ed. Washington, DC: ASM Press.

Food and Drug Administration. 1987. *Guidelines on sterile drug products produced by aseptic processing.* Washington, DC: U.S. Food and Drug Administration.

General Services Administration. 1991. *Current good manufacturing practices.* Title 21, Code of Federal Regulations, Part 211. Washington, DC: General Services Administration.

General Services Administration. 1992. *Airborne particulate cleanliness classes in cleanrooms and clean zones.* Federal Standard 209E. Washington, DC: General Services Administration.

Groves, M., W. Olson, and M. H. Anisfeld, eds. 1991. *Sterile pharmaceutical manufacturing: Applications for the 1990s.* Buffalo Grove, IL: Interpharm Press.

National Aeronautics and Space Administration. 1967. *Standards for clean rooms and work stations for the microbially controlled environment.* Pub. NHB 5340.2. Washington, DC: National Aeronautics and Space Administration.

National Institutes of Health. 1994. Guidelines for research involving recombinant DNA molecules. *Fed. Regist.* 59:34496–34547.

National Sanitation Foundation. 1976. *Class II (laminar flow) biohazard cabinetry.* Standard #49. Ann Arbor, MI: National Sanitation Foundation.

Chapter 5

Andersen, A. A. 1958. New sampler for the collection, sizing, and enumeration of viable airborne particles. *Journ. Bacteriol.* 76:471–484.

Cox, C. S. 1987. *The aerobiological pathway of microorganisms.* New York: John Wiley and Sons.

Index

dry heat, 68, 73, 75, 76–78, 90,
    91, 135
gaseous, 84–85, 91
heat, 100
level of assurance of, 69, 70
moist heat, 68, 73–75, 76, 91,
    135, 137
radiation, 91, 100
sterilants for, 82, 84–85
terminal, 100–101, 132, 140
time required for, 68, 70, 71, 72,
    75, 76–77, 84
validation of, 90–92
strain, 4
streptoccus, 7
Streptococcus, 7
Streptococcus lactis, 50
subspecies, 4
surface sampling, 126–128
surfactant, 86

taxonomy, 4–5
terminal sterilization. *See* sterilization
thermal death, 69
thermal death time, 72, 78
thermocouple, 90
thermophile, 41, 42, 140
thioglycollate broth, 43
titer, 63
tobacco mosaic virus, 25
toluene, 116
touch plate, 128
toxicity, 82, 85, 86, 87
training, 111
transduction, 17, 18
transformation, 17
translation, 17
transmission electron microscope,
    14. *See also* microscopy
Treponema pallidum, 50
tryptic soy medium, 39
tuberculosis bacilli, 86, 89
tunneling electron microscope, 35.
    See also microscopy
turbidity, 41, 58, 61

turbidometry, 58, 59
Tween 80, 127

ULPA filter, 100
ultrafiltration, 11, 93
Ultra Low Penetration Air filter. *See*
    ULPA filter
ultrasonic bath, 125
unidirectional flow, 97, 98, 99, 101, 111
use dilution test, 89–90, 140

vaccines, 3, 63
validation, 92, 140
vector flow, 98
vegetative cell, 140. *See also* spore
vertical flow, 98, 125
viable (viability), 68, 96, 117, 136,
    140, 141
    biological indicators and, 91
    contamination, 95, 100, 111, 116,
        127, 128
    impinger assay of, 124
viable population, 47, 51, 52, 57,
    60, 69, 71, 94, 129
vibrio, 5, 6
viricide, 85, 141
virion, 24, 25, 26, 63
viruses, 1, 3, 18, 24–28, 100
    counting, 62–63
    elimination of, 81
    life cycle of, 27
    shapes of, 25
    size of, 26, 30
viscosity, 128
volatility, 87

water for injection, 93
wavelength, 28–30, 31, 57, 59, 78
wet heat, 68
witness strip, 125–126
Woods lamps, 79

yeasts, 3, 19, 20–22, 43, 62